Lilia BOUSSOUF
Imene Mesbout
Meriem Mestar

Fresh figs (Ficus carica L.): physicochemical and nutritional quality, antioxidant activity.

Lilia BOUSSOUF
Imene Mesbout
Meriem Mestar

Fresh figs (Ficus carica L.): physicochemical and nutritional quality, antioxidant activity.

Fresh figs: quality and antioxidant activity

Imprint

Any brand names and product names mentioned in this book are subject to trademark, brand or patent protection and are trademarks or registered trademarks of their respective holders. The use of brand names, product names, common names, trade names, product descriptions etc. even without a particular marking in this work is in no way to be construed to mean that such names may be regarded as unrestricted in respect of trademark and brand protection legislation and could thus be used by anyone.

Cover image: www.ingimage.com

This book is a translation from the original published under ISBN 978-620-3-45920-3.

Publisher:
Sciencia Scripts
is a trademark of
Dodo Books Indian Ocean Ltd. and OmniScriptum S.R.L publishing group

120 High Road, East Finchley, London, N2 9ED, United Kingdom
Str. Armeneasca 28/1, office 1, Chisinau MD-2012, Republic of Moldova, Europe
Printed at: see last page
ISBN: 978-620-6-09682-5

THANKS

It is with great pleasure that we reserve these lines as a token of our gratitude and profound appreciation to all those who, from near and far, have contributed to the realization and completion of this work.Above all, we thank Almighty God for helping us to complete this modest work. We would like to express our sincere thanks to our sponsor, Ms BOUSSOUF Lilia, PhD, Department of Applied Microbiology and Food Science, Faculty of Natural and Life Sciences, Mohamed Seddik Benyahia University, Jijel, for accepting the scientific direction of our final year project. May she be reassured of our deep gratitude.

We would like to thank :

Mr. BOUBZARI Mohamed Tahar, Doctor, Mohamed Seddik Benyahia University, Jijel, for the great privilege of chairing our defense jury. May he be assured of our respectful consideration.Mr. Dairi Soufian, Doctor, Mohamed Seddik Benyahia University, Jijel, for agreeing to spend time examining and judging this work.

To our dearest parents,

Thank you for being there for us, for your support during our studies and for everything you've given us to get here. We could never repay you. We love you all We'd also like to thank our dear friends Chouaib, Loukmen, Fatima, Farida, Zahra, Dalila and Fatiha for their support, kindness and sympathy, and wish them all the best for the future.

CONTENTS

INTRODUCTION

Figs, a product of fruit arboriculture, are the fruit of the fig tree, a member of the Moraceae family, which is emblematic of the Mediterranean basin, where it has been cultivated for thousands of years (Slatnar et al., 2011). It is generally eaten fresh or dried (Faleh et al., 2012; Stalin et al., 2012).

In the northern Mediterranean region, fig trees produce one or two crops per year, depending on the cultivar (Veberic et al., 2008). Every year, over one million tonnes of fresh figs are harvested from 308,460 hectares worldwide. Mediterranean countries are the main fig producers. Algeria produces 12.54% of the world's total fig crop (FAO, 2016).

In Algeria, fig trees are grown all over the country (coastal, steppic and Saharan zones), due to its pedoclimatic adaptation, its nutritional and therapeutic values, and its place in the culinary practices of Algerians. Algerian fig production is concentrated mainly in the mountainous area of Kabylie (Bejaia and Tizi-Ouzou account for 27% and 13% of total national production respectively) (MADR, 2012).These fruits have shapes, colors, tastes, technological and therapeutic properties that differ between varieties and are generally named according to their shape, color and the region in which they are most commonly grown. Recently, the Institut Technique de l'Arboriculture Fruitière in Algeria described 40 varieties, including edible varieties and caprifigue types (Chouaki, 2006; Meziant et al., 2015).Figs are an excellent source of energy and nutrients, thanks to their high sugar, fiber and mineral content (Imran et al., 2011). They are fat- and cholesterol-free and contain numerous bioactive substances such as vitamins (vitamins C and E) (Guvenc et al., 2009) and phenolic compounds with high antioxidant potential (Slatnar et al., 2011) including flavonoids and anthocyanins (Vinson, 1999).It is well known that the quality of fresh fruit, including figs, is determined not only by its nutritional and bioactive composition, but also by other parameters linked to sensory characteristics, notably firmness, visual appearance, taste and aroma. Recently, a great deal of research has focused on the nutritional, physicochemical and pharmacological quality of this fruit.Our dissertation is a literature review covering four chapters:

• The first chapter deals with the agronomic aspects of the fig tree.

• The second chapter will be devoted to the fruit (description and morphology, classification, reproduction and maturity, composition and nutritional value, therapeutic properties, fig production and technology).

• The third chapter illustrates figs' polyphenolic profile and antioxidant power.

• The fourth chapter will be devoted to previous studies characterizing the physicochemical and nutritional quality, polyphenol composition and antioxidant activity of figs.

I.3.2. The female fig tree

Female fig trees, or domestic fig trees, produce the best edible figs. The flowers that make up the fruit are only long-styled female flowers (longistylus), which, once fertilized by caprifigus pollen, produce achenes (Vidaud, 1997). The exception is certain figs that are eaten and ripen early in the summer. These are "flower" figs, or "El Bakor", which are made up of female flowers that have not been fertilized due to a lack of pollen, and are therefore seedless, but still manage to ripen by "parthenocarpy" (from parthenos and carpon, meaning virgin and fruit, respectively) (Figure 02) (Garrone et al., 1998).

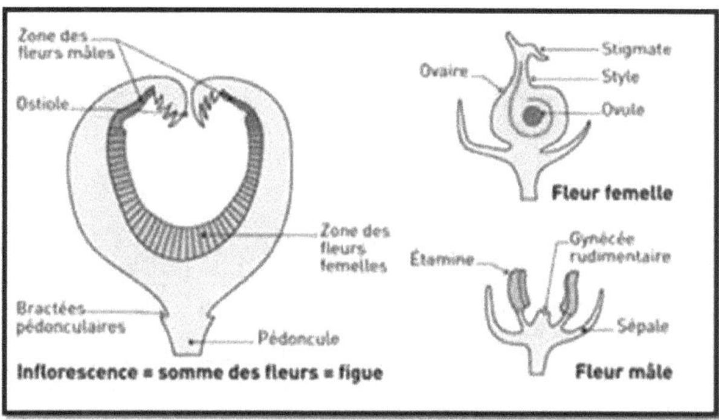

Figure 02: Difference between female and male fig flowers (Vidaud, 1997).

I.3.3. The blastophage

The pollinator of Ficus carica is a tiny insect, a Hymenoptera of the order Chalcoideae and family Agaonideae: the Blastophagus or Blastophaga psenes L. The fig tree can only be pollinated naturally by the Blastophagus, and the latter cannot reproduce outside the fig tree's fruiting bodies: they have formed a true symbiotic unit.Each Ficus species has a specific pollinator, belonging to the same Agaonideae family; there are as many pollinator species as there are Ficus species. In the blastophagus, sexual dimorphism is very pronounced; the male and female are morphologically different (Figure 03). The insect responsible for pollination is the female. It's about 2 mm long, black, winged, equipped with an ovipositor that enables it to lay eggs, and roughly the same length as the styles of brevistylated female caprifig flowers. The male is smaller than the female, yellowish and wingless (Garrone et al., 1998).

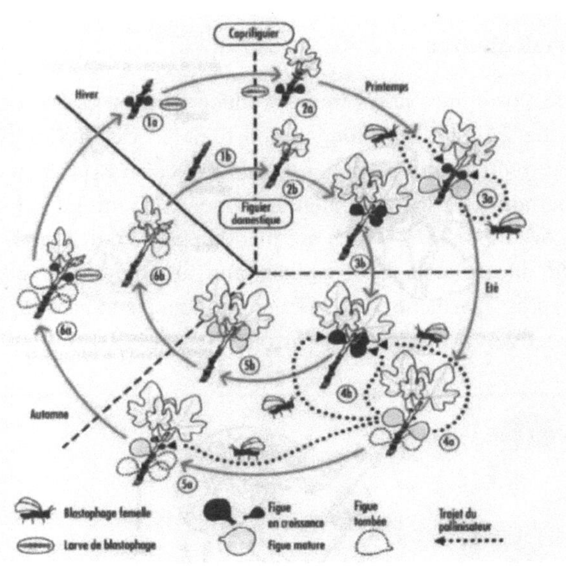

Figure 03: Simplified life cycle of the fig tree and its pollinator (Vidaud, 1997).

I.4. Description of fig tree

Ficus carica L, or common fig, is a fruit tree in the Moraceae family of the ficus genus, whose scientific name is Ficus carica L. It is considered the emblem of the Mediterranean basin, where it has been cultivated for thousands of years. Our ancestors used the various parts of this tree - leaves, latex, bark and roots - for medicinal purposes. The Ficus genus has an average of 850 species, and Ficus is probably the leader of all plant genera given the number of species it has (Lansky and Helena, 2011). It is a monoecious shrub of variable size, with deciduous or large leaves. It reaches heights of ten to twelve meters, with smooth gray bark. Its fragrant leaves are 12 to 25 centimeters long and of
10 to 18 centimetres in diameter and deeply lobed with three or five lobes, with numerous branches and spreading trunks rarely over 7 feet in diameter (Ahmad et al., 2013). The flowers are grouped into distinctive inflorescences called sycones, which yield the fig fruit (Figure 04) (Bayer et al., 2005). The plant's latex is milky white and contains mainly ficin, a protein hydrolytic enzyme. The plant's root system is generally shallow and spreading (Badgujar et al., 2014).

Figure 04: Fruit-bearing shoots of Ficus carica L. (Bakshi et al., 1999).

I.5. Morphology of fig tree

I.5.1. The buds

At the end of the entire stem is a terminal bud, which contains all the elements of the future stem, as well as the terminal meristem, which produces new plant parts (Vidaud, 1997).

In the axil of each leaf, or the scar it leaves after falling off, there is a lateral or axillary bud. Closer examination reveals two buds: one pointed and vegetative, the other rounded and floral. Sometimes, two rounded buds are present on either side of the vegetative bud (Vidaud, 1997).

I.5.2. The inflorescence

The fig or sycone is a group of flowers in the same structure. The flowers are not visible from the outside. They are enclosed in a kind of urn, with a hole, the ostiole, which opens away from the short stalk bearing the fig (Vidaud, 1997).

I.5.3. The flowers

Fig flowers can be either male or female. Depending on their distribution, the individual bearing them is either male or made up entirely of female flowers. In male figs, both female and male flowers are present. The latter are few in number and located all around the ostiole (Vidaud, 1997).

I.5.4. The fruit

The fig is a round fruit weighing between 30 and 65g, depending on the variety. It consists of a colored outer skin and an inner part containing a liquid called latex, rich in protease and lipase, which accounts for 10-20% of the fruit's weight (Ouaouich et al., 2005).

I.6. Development cycle

The description of the biological cycle begins in winter, when the fig and the insect (cycle 1a, 1b) are at rest. The cycle doesn't start up again until April, with the emergence of new fig shoots (cycle 2a, 2b) and the resumption of blastophagus larval development (cycle 2a). Adult females emerge in May without being loaded with pollen, as the male flowers of the caprifiguier have no pollen (cycle 3a). The new generation of blastophagus reaches maturity in mid-July with the emergence of pollen-laden females (cycle 4a). The insect is attracted by a receptive fig on the same tree, or on a different tree. different tree (domestic fig tree) (cycle 4b). The insect deposits pollen, enabling the ovule to be fertilized and develop into a seed. These figs are the future edible autumn fruit (Kjellberget et al., 1988). The females that emerge later (early August), enter a caprifigus fig (cycle 5a), then receptive, and lay their eggs without difficulty and the female fig trees see their autumn figs mature (cycle 5b). The larvae then begin to develop (cycle 6a), but winter blocks their development and a new cycle can begin (Caraglio, 2005).

I.7. Agro-ecology of the fig tree

The fig tree is considered a subtropical plant, or for temperate to warm climates, but it can be grown at altitude (up to 1500 m) in many tropical countries (Leroy, 1968). It is distributed in south-western Asia and the Mediterranean, from Turkey in the east to Portugal in the west, via northern Africa. It is also grown in the USA (California), Chile, Arabia, Persia, India, China and Japan (Chawla et al., 2012).Of all plants, the fig tree appears to be the least demanding. It can grow in the most rocky soils in the presence of a minimum of water (Valizadeh et al., 1987). The rich nutritional content of the tissues enveloping its fruit attracts consumers (mammals and birds) capable of distributing its seeds for miles around. It is thus the primary agent in the formation of arable land in arid zones (Garrone et al., 1998). Fig trees deserve more attention: they require irrigation in hot weather and deep, well-drained soil, moderately fertilized and rich in organic matter. Heavy, moist soil tends to encourage growth over production (Lim, 2012). According to Oukabli (2003), the fig tree's actual annual water requirements are estimated at 600 mm supplied mainly in spring and early summer. According to (Walali et al., 2003), the optimum soil pH varies between 6 and 7.7. The fig tree can tolerate the salinity of coastal areas and freezing winter periods; according to Stover et al. (2007), it can withstand -10°C. Pruning in spring, when sap rises, helps control tree size and increases production (Lim, 2012).The fig tree is propagated either by wood cuttings, which are taken in February during dormancy (vegetative rest), or by herbaceous cuttings, which are taken in June-July with the year's green shoots, or by grafting if a change of variety is envisaged (Brien and Hardy, 2002). The distance between plantations varies according to soil richness, annual rainfall and irrigation possibilities. It is 3 to 6 m along the row and 5 to 7 m between rows (i.e. 250 to 400 plants per hectare). Fruit set begins from from the 3rd year, but the maximum yield (5 tons/ha in dry land to over 20 tons/ha in irrigated cultivation) is reached after 6 years (Oukabli, 2003).

I.8. Geographical distribution of fig

I.8.1. In the world

Figs were one of the first fruits cultivated in the Mediterranean region, and a wild genetic resource for fig species that still exist in many countries. Syria and Anatolia are the fig tree's natural habitats, and from there it was transferred to North Africa, Spain, Mexico, Chile, Peru and California. It has also been transported to South America via France and to Mesopotamia, Iran and India from Anatolia (Preedy and Watson, 2014). The Middle East is the world's leading producer (notably Turkey). The fig tree has long been found throughout the Mediterranean Basin, from Syria to Morocco and from Turkey to Portugal. Over the centuries, the fig tree has been introduced on every continent (South Africa, Australia, and especially North and South America by Spanish colonists). The fig tree is grown wherever the climate is similar to that of the Mediterranean (Vidaud, 1997). The common fig, F. carica, was one of the first cultivated fruit species and has become an important crop worldwide (Shamin-Shazwan et al., 2019).

I.8.2. In Algeria

The fig tree is one of Algeria's three main fruit crops: olive, fig and citrus. The vast majority of plantations are located in Kabylia (Chouaki et al., 2006). The north-central region of Algeria encompasses the majority of fig trees, with over 61% of the national orchard, followed by the eastern highlands region with over 19%. The North-West and North-East regions account for over 11% (Table 01). Fig trees are concentrated in the Kabylie wilayates, namely Béjaia with 13,352 ha or 28.65% of the orchard, followed by the wilaya of Tizi-Ouzou with 6,387 ha or 13.70%. These two wilayates alone account for over 40% of the fig orchard, followed by Sétif with 4,922 ha or 10.56%, Borj Bou Arredj with 2,033 ha or 4.36% and Bouira with 1,928 ha or 4.31%. Total number of trees is 6,044,550, of which 4,719,950 in mass and 1,324,600 isolated with 4,610,040 fig trees in connection. Plantations are mostly located in mountainous areas, on poor quality soils with a high stony load. The fig tree occupies a fragmented space in small, uneven plots (ITAFV, 2003).

Table 01: Distribution of tree species in Algeria by region (excluding vines and date palms) (MADR, 2005).

Regions	Olive trees (ha)	Citrus (ha)	Fig trees (ha)	Core species and/or Pome fruit (ha)	Total area (ha)
MOSTAGANEM	4 671	4 079	1 085	7 687	17 522
ORAN	3 962	674	680	5 370	10 686
A.TEMOUCHENT	2 875	483	934	5 772	10 064
TLEMCEN	4 760	2 446	384	18 257	25 847
BASCRA	9 977	4 232	580	6 527	21 316
RELIZANE	6 262	4 417	460	4 976	16 115
North West	36 422	16 338	4 385	56 246	113 391
BOUIRA	18 835	421	1 928	5 223	26 407
BEJAIA	50 663	1 890	13 352	5 962	71 867
TIZI-OUZOU	32 443	1 349	6 378	4 768	44 952
BOUMERDES	6 005	2197	1 155	4 657	14 014
ALGER	19	5 065	32	4 171	9 287
BLIDA	2 411	15 809	845	9 200	28 265
North Central	123 567	38 486	28 486	85 879	276 418
JIJEL	10 520	415	142	2 806	13 883
SKIKDA	5 758	2 214	245	4 236	12 453
MILA	4 274	0	190	1 901	6 365
GUELMA	7 143	835	172	4 718	12 868
EL-TARF	1 982	2 127	141	2 082	6 332
SOUK-AHRAS	1 392	0	118	2 849	4 359
North East	31 871	6 113	1 109	21 205	60 298
TIARET	1 573	0	836	14 434	16 843
SAIDA	2 732	0	338	3 800	6 870
NAAMA	552	2	82	1 763	2 399
EL-BAYADH	405	0	750	4 492	5 647
High Plateaux West	5 262	2	2 006	24 489	31 759
MSILA	1 755	30	868	11 398	14 051
DJELFA	2 807	0	144	8 722	11 673
LAGHOUAT	162	26	96	4 110	4394
Central Highlands	4 724	56	1 108	24 230	30 118

CETIF	12 865	0	4 922	6 617	24 404
B.B.ARREDJ	15 947	0	2 033	4 306	22 286
BATNA	3 526	0	320	9 941	13 787
O.E.BOUAGHI	142	0	66	1 203	1 411
TEBESSA	2 006	0	824	5 553	8 383
KHENCHELA	1 440	5	1 034	10 940	13 425
High Plateaux East	35 926	5	9 199	38 566	21
BECHAR	21	22	45	158	246
TINDOUF	7	0	35	35	77
ADRAR	0	0	0	0	0

I.9. Problems associated with growing fig trees

Nowadays, the fig sector is faced with a number of adversities:

► The abandonment of orchards, as well as urbanization and the development of road and water infrastructures (installation of several dams), have also been the main causes of the disappearance of certain cultivars and the aging of trees (Chouaki et al., 2006).

► The planting of other fruit crops such as pomegranates, plums and peaches in fig-growing areas (Arpaci, 2015).

► Fig trees thrive in areas with low humidity, plenty of sunshine and hot, dry summers. When young, growing shoots can be damaged (at -1°C) (Walali et al., 2003).

► Another factor negatively affecting fig production occurred with the potential increase in irrigation (Arpaci, 2015).

► The fig tree adapts to a wide range of soils, from heavy clay soils to the most difficult. sandy soils, but fears high concentrations of sodium and boron (Walali et al ., 2003).

► In poorly maintained orchards (no pruning), the fig mealybug (Lepidosaphes ulmi) develops on bark, leaves and fruit, secreting a pinkish-white waxy substance. Other common diseases include Aspergillus niger and botrytis cinerea. Insects include the fig psyllid, which attacks leaves and young shoots, and the black fig fly, which can cause 60 to 70% of fruit to drop in certain years (Walali et al., 2003).

II.1. Description and morphology

The fig "fruit" is a composite formed by a hollow shell of receptacle tissue enclosing hundreds of individual pedicellate drupelets that develop from the individual female flowers lining the receptacle wall, with a small scaled opening (called the ostiole or eye) at the distal end. The tiny flowers and even the initial prosyconium are so small that figs were once thought to bear fruit without ever forming flowers. This composite fruit is called a "syconium". The ripe fruit of the edible fig has a somewhat hard skin, a whitish inner rind and a soft gelatinous pulp made up of individual ripe drupelets (Figure 05). Seeds in the drupelets range from practically non-existent to subtly crunchy (Stover et al., 2007). The color of figs varies from dark purple to green; at maturity, the fruit has a foreign dark purple-violet color, reaching around 7.5 cm in length and weighing between 60 and 90 g. Fig pulp is pink-red in color, with a central cavity (Silva et al., 2009). There are over 700 varieties of fig. The most common are the black fig (sweet and rather dry), the green fig (juicy and thin-skinned) and the purple fig (sweeter, juicier, more fragile) (Haesslein and Oreiller, 2008).

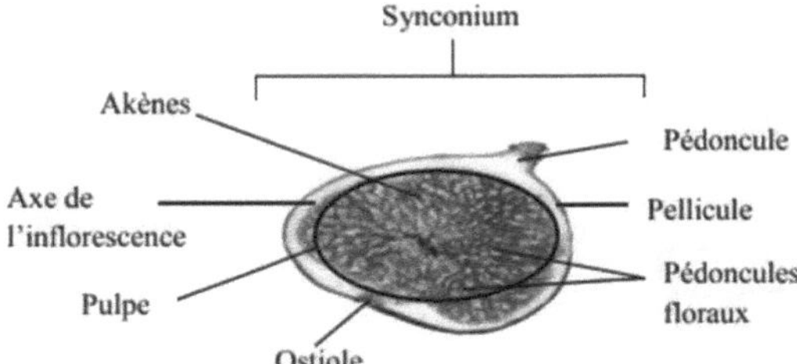

Figure 05: Morphological characteristics of figs (Haesslein and Oreiller, 2008).

II.2. Classification of figs

Four types of fig are described according to crop and pollination characteristics.

II.2.1. Persistent figs

This is the type known as the common fig (Flaishman et al., 2008). These figs develop parthenocarpically (without fertilization) and do not require the stimulation of pollination to ripen the sycone (Condit, 1955).

II.2.2. Deciduous figs (type Smyrna)

Smyrna-type figs, which produce a single main crop, have a maximum starch concentration in early spring, midsummer and late autumn (Flaishman et al., 2008). Varieties of this type produce fruit filled with true achenes. Blastophages and caprifigues are necessary for normal fig development. If pollination does not take place, the fruit falls from the tree. Figs of this type are generally sold in dried form (Lim, 2012).

II.2.3. Intermediate figs (San Pedro type)

Fig trees in this category produce two crops in a single season: the first crop, called Breba (flowering figs), is borne on the previous year's growth unit; the figs are parthenocarpic and do not require pollination. The fruits of the second harvest develop on the current year's growth unit; the figs are Smyrna-type and require pollination (Lim, 2012).

II.2.4. caprifigues

Figs of this type are not edible, but produce the pollen essential for the production of edible figs of the Smyrna and San Pedro types (Lim, 2012).

II.3. Reproduction and maturity

The reproductive system of Ficus species is unique. Each Ficus species has an associated agaonid wasp species. Ficus species can only be pollinated by their associated agaonid wasps and, in turn, the wasps can only lay eggs in their Ficus fruit. For successful pollination and reproduction of Ficus species, the associated pollinating wasp must be present. Conversely, for successful reproduction of agaonid wasps, their associated Ficus species must be present. The pollinating wasp of F. carica is Blastophaga psenses (L.) (Starr, 2003). The fig ripening process is classified as climacteric, showing an increase in respiration rate and ethylene production early in the ripening phase (Marei and Crane, 1971) (Figure 06). Fruit development reveals a double sigmoid curve. Fruit diameter increases rapidly during the first growth period, but weight increases slowly. There is almost no change in fruit diameter or weight during the second growth period. The diameter and weight of both dried and cooked figs increase rapidly during the third period. Phase III is a result of cell expansion, and is also the ripening phase, which includes the change in color. 7% of dry weight and 90% of total sugars are accumulated during this growth phase, which lasts from 2 to 5 weeks in most varieties (Åkesson, 2010) (Figure 07).

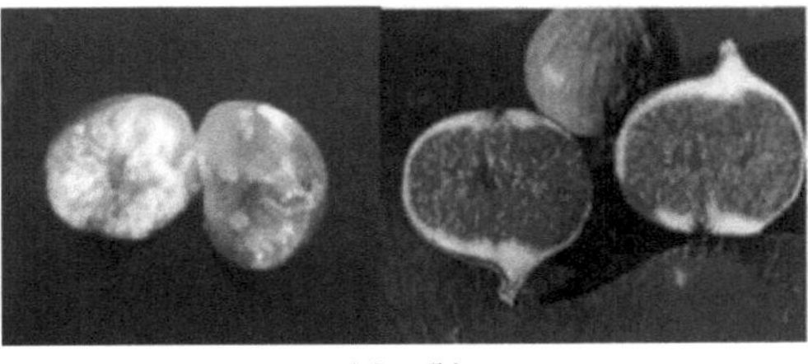

(a) (b)

Figure 06: (a) immature fig; (b) mature fig (Afriyanti et al., 2018).

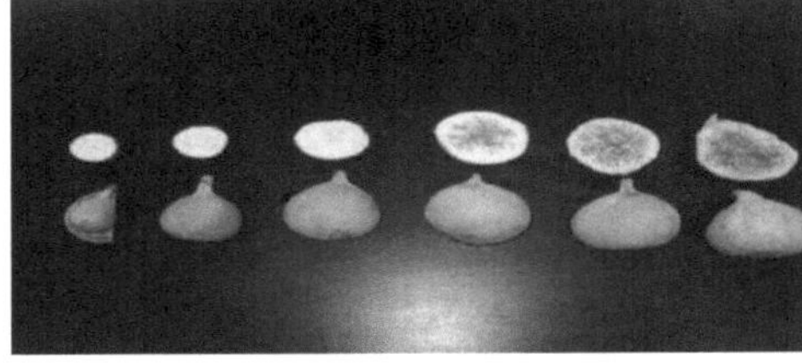

Figure 07: The different stages of fruit development, inside and out

(Crisisto et al., 2011).

II.4. Composition and nutritional value of figs

Figs are an excellent source of minerals, vitamins and dietary fiber; they are fatty and cholesterol-free and contain a high number of amino acids. As with other fruit species, figs contain sugars and organic acids that influence their quality (Mahmoudi et al., 2018).

II.4.1. The carbohydrates

Fig fruits have a high commercial value and are an important supplement to human nutrition, providing energy compounds in the form of starches and sugars such as glucose and fructose (Mendoza-Castillo et al., 2019). Qualitative analysis of fig carbohydrates revealed the presence of free sugars mainly glucose, fructose, a small amount of sucrose and traces of galactose and arabinose (Golubev et al., 1987). According to Omondi Owino et al. (2004), figs also contain xylose. The low sucrose concentration in figs could be the result of anabolic processes and respiration during fruit development (Veberic and Mikulic- Petkovsek, 2016). In the fresh state, the total sugar content varies from 9.8 to 18.9% (Babazadeh darazi, 2011). Data put forward by Lim (2012), show that 100 g of fresh figs contain 19.18 g of total carbohydrates, including 16.26 g of sugars. The same portion of dried figs provides 63.87 g of carbohydrates, including 47.92 g of sugars (24.79 g of glucose, 22.93 g of fructose, 5.07 g of starch, 0.13 g of galactose and 0.07 g of sucrose).

II.4.2. Proteins and amino acids

According to Lim (2012), fresh figs contain 0.75 g/100 g of protein. It contains around 17 amino acids, including aspartic and glutamic acids (Solomon et al., 2006). According to the study carried out by Lim (2012), the content of "acid" amino acids is higher than that of other amino acids contained in figs, whether fresh or dried (Table 02).

Table 02: Amino acid composition of fresh and dried figs (Lim, 2012).

Amino acid	Fresh fig (mg/100g)	Dried fig (mg/100g)
Aspartic acid	176	645
Glutamic acid	72	295
Alanine	45	134
Arginine	17	77
Cystine	12	36
Glycine	25	108
Histidine	11	37
Isoleucine	23	89
Leucine	33	128
Lysine	30	88
Methionine	6	34
Phenylalanine	18	76
Proline	49	610
Serine	37	128
Threonine	24	85
Tryptophan	6	20
Tyrosine	32	41
Valine	28	122
Total	0.64 g/100 g	2.75 g/100 g

II.4.3. The lipids

Figs contain a small amount of lipids (around 1.9%). Despite their low content, lipids have a fundamental influence on storage life, organoleptic properties and the nutritional and biological value of figs (Kolesnik et al., 1987). The Fig lipids are characterized by a high level of unsaturation (>68%) of monovalent fatty acids, the majority of which are polyunsaturated and which, in certain cases, may be responsible for the oxidative deterioration of figs and their derivatives. Neutral lipids represent the largest fraction of total lipids, with tri-acylglycerol the predominant compound at 50%; sterol esters, fatty acid esters and free sterols are also present in considerable quantities. Phospholipids represent only a small fraction (Kolesnik et al., 1987).

II.4.4. Fiber

Figs are fibrous plants (Guvenc et al., 2009). Dietary fibers include lignin and carbohydrates such as cellulose, hemicelluloses, pectins, resistant starches and non-digestible oligosaccharides. They are divided into two groups according to their solubility in water: soluble, viscous and fermentable fibers and insoluble, non-viscous and slowly fermentable fibers (Ramulu and Rao, 2003). The study carried out by Ramulu and Rao (2003) showed that fresh figs contain 5g of dietary fibre per 100g of fresh matter, including 2.6g of insoluble fibre and 2.4g of soluble fibre (48% of total fibre). Thanks to its high fiber content, figs are an excellent remedy for constipation (Bidri, 2018). Its fiber is highly effective in stimulating the intestines (Du Toit et al., 2001).

II.4.5. The vitamins

Figs are rich in water-soluble vitamins B1, B2 and C (Farahnaky et al., 2009). Fat-soluble vitamins are also present in figs, with vitamins E and K predominating (Lim, 2012). According to Pande and Akoh (2010), the vitamin E content of figs is 0.3mg/100g for its γ form, 0.2mg/100g DM for its α form and traces for the β form. According to Guvenc et al, (2009) the whole fruit contains vitamins K1 (4.05 µg/100 g), D2 (0.2 µg/100 g) and D3 (3.57 µg/100 g), α-tocopherol (0.35 µg/100 g), γ-tocopherol (0.9 µg/100 g) and δ-tocopherol (0.20 µg/100 g). The dried fruits of Ficus carica have been reported as an important source of vitamins (Mawa et al., 2013).

II.4.6. minerals

Figs contain an ash content of 0.66 g/100 g fresh figs (Lim, 2012). In addition, it contains significant quantities of minerals necessary for metabolism, namely P, K, Ca, Mg, Na, Fe and Z (Mendoza-Castillo et al., 2019) (Table 03) . In addition to minerals Favier et al (1993) reported the presence of iodine in figs at a level of 1.5 µg/100 g of fresh fig. They also noted the presence of fluoride (20 µg/100 g).

Table 03: Mineral composition of fresh figs (Lim, 2012) .

Constituent	Fig (mg/100g)
Posssium (K)	232
Calcium (Ca)	35
Phosphora (P)	14
Magnesium (Mg)	17
Sodium (Na)	1
Iron (Fe)	0.37
Zinc (Zn)	0.15
Copper (Cu)	0.07
Manganese (Mn)	0.128
Selenium (Se)	0.2 µg /100g

II.4.7. Organic acids

Figs are very rich in organic acids such as oxalic, malic, citric, shikimic and fumaric acid (Oliveira et al., 2009). In a study by Pande and Akoh (2010), fresh figs were found to contain oxalic acid (17.9 mg/100 g), ascorbic acid (14.2 mg/100 g) and succinic acid (10.2 mg/100 g), in addition to other organic acids.

II.5. Therapeutic properties of fig

Edible figs, Ficus carica L., play an important role in human and animal nutrition worldwide. Humans have consumed the fruits of these trees since the earliest times, and have used them and other parts of the tree for medicinal purposes.Regular consumption of figs helps to lower high blood pressure, control cholesterol, relieve constipation, prevent colon cancer, control blood glucose levels, prevent the onset of type 2 diabetes and is an important drainer for the respiratory and intestinal tracts. Recently, the pharmacological aspects of Ficus carica have been specifically examined and have shown that this plant (Zhang and Jiang, 2006) :

ﺢ Has antitumor effects, reducing toxicity and side effects in actinotherapy and chemotherapy.

ﺢ Controls hyperglycemia and hyperlipidemia, and boosts resistance to oxidation.

¿Acts against bacteria and viruses.

Figs' high fiber and potassium content helps prevent many ailments. Fiber improves digestion, allowing nutrients to be better absorbed and controlling the sugar level absorbed during a meal, preventing it from rising too quickly. Fiber also protects against many intestinal ailments, and studies show that a high fiber intake protects against cardiovascular disease. As the Western diet is generally far too rich in sodium and promotes heart disease, a good way to counteract the effects of high sodium consumption is to increase potassium intake, which is found in good quantities in figs. Figs contain a special enzyme known as the fig enzyme or "ficin", which has been shown to play an important role in digestion. It has also been shown to be beneficial for milk flow, hemorrhoids, chronic constipation, gout, lung disease, premenstrual pain, epilepsy, mouth ulcers, gingivitis, tonsillitis, pharyngitis, vitiligo, wound healing and various ulcers (Lansky and Helena, 2011).

II.6. Fig production

II.6.1. Production worldwide

Worldwide fig production represents over one million tonnes (FAO, 2016) (Table 04). This production is largely based on dry figs, which are hardier and more storable, mainly dominated by Mediterranean countries. The major producers are Turkey with 29.16% of world production, followed by Egypt with 16% and Algeria with 12.58% (FAO, 2016).

Table 04: World fig production (Tonnes) (FAO, 2016).

Country	Production in Tons
Turkey	305450
Egypt	167622
Algeria	131798
Iran	70178
Morocco	59881
Syria	43098
United States	31600
Brazil	26910
Spain	25224
Tunisia	22500

II.6.2. Algerian production

Fig production in Algeria, especially in the dry state, is as important as date and citrus production. Fig trees are found in small plantations all over northern Algeria: in Oran, near Mostaganem, Mascara and Constantine, but 80% of producing trees are concentrated in the Tizi-Ouzou and Bejaïa regions. For this reason, the Kabyle fig groves form the backbone of Algerian production (Anonymous, 2005). Fig tree cultivation is specifically developed in the Kabylie region, where favorable socio-cultural conditions prevail. The climatic factor, characterized by high autumn rainfall, represents a serious problem for drying stations, which are generally traditional.

II.7. Technology fig

II.7.1. Processing and use of fig

Figs can be put to a variety of traditional and industrial uses. They are widely consumed both fresh and dried. Fresh figs are highly perishable, which is why they are mostly dried or canned. They can also be used to make many food products, such as jam and candied fruit, alcoholic beverages, etc. Peeled or unpeeled, fruit can be baked in a variety of ways, such as in pies, puddings, cakes, bread or other baked goods, or added to mixed ice cream. Owners preserve whole fruits in sugar syrup or prepare them in jam, marmalade, or paste. Fig paste (with wheat and add corn flour, whey, syrup, oils and other ingredients) forms the filling of the baked product (Chawla et al., 2012). Fig leaves can also be used as livestock feed. The latex, dried and powdered, is used to coagulate milk. Fig cell cultures are also valued as a source of protease. Several other medicinal uses of fig products have been reported (Oukabli, 2003).

II.7.2.1. Dry figs

Figs (fruits of Ficus carica L.) are generally dried in the traditional way, without any treatment using driers or by exposure to the sun (Haesslein and Oreiller, 2008). Figs intended for drying must acquire a pasty consistency on the tree. The skin is not only split, but wrinkled and withered. Figs are harvested in cool weather with humidity levels of 40-60% (Ait Haddou, 2014). The two main pre-treatments needed to dry figs are blanching and sulfiting. Bleaching is a heat treatment designed to destroy enzymes that can spoil figs. It affects drying speed (Ferradji et al., 2011). According to Ouauiche et chimi (2005), raw materials intended for drying always undergo preliminary preparation (cleaning, sorting, grading, bleaching/fumigation, etc.) with a view to subsequent processing. These preparation operations vary according to the nature of the raw material and the product to be obtained. The main ones are shown in the diagram below (Ouauiche and Chimi, 2005) (Figure 08).

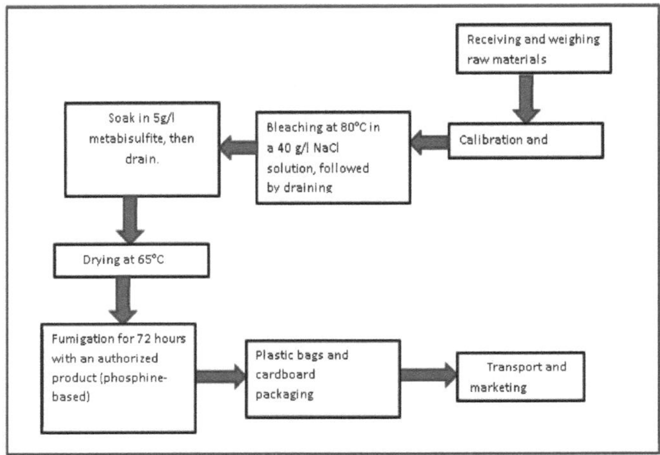

Figure 08: Diagram of dried fig production (Ouauiche and Chimi, 2005).

II.7.2.2. Fig jam

Jam plays a very important role in the human diet, especially in the first meals of the day. This is due to its composition and fruit content. Its high sugar content accounts for the lion's share of its energy value. The fruit provides 10-15% crude cellulose, mineral elements, pectic matter and trace organic acids. 100 grams of jam contains 260 to 285 calories (Leroux and Schuber, 1983). Temperature and storage time are among the most important factors influencing the quality of food products. The optimum storage temperature for jams is around 4°C, to avoid the degradation of certain components (Amora et al., 2012). Storage and even cooking temperatures must therefore be carefully controlled to preserve jam quality. This is the background to the work carried out by Patil et al, (2017) which focuses on the standardization of the recipe for making fig jam to assess its quality and study the impact of temperature on its preservation. To do this, the authors followed the steps illustrated in figure 09 for making fig jam.

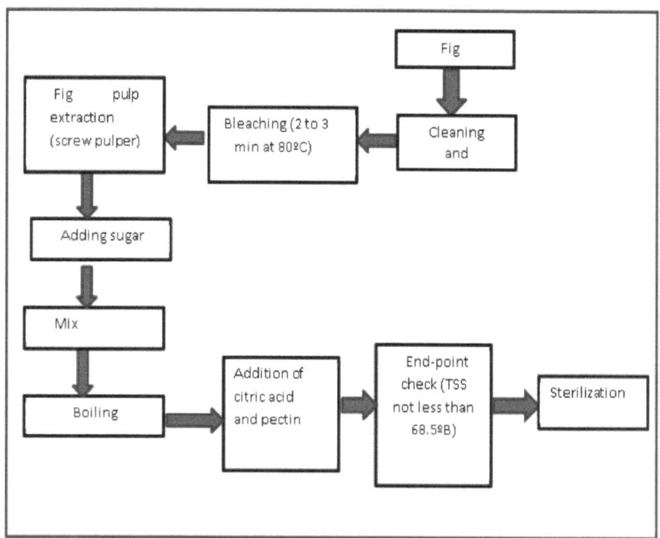

Figure 09: Diagram of fig jam production (Patil et al., 2017).

II.7.2.3. Fig syrup

Fig syrup is a fruit concentrate used as a common ingredient in the preparation of typical foods and in cakes in particular (Puoci et al., 2011).

Gabriele et al, (2010) in their study dedicated to analyzing the influence of parameters such as pH, temperature.... on product rheology, prepared a fig syrup by mixing dried figs with boiling distilled water. This first step enables the extraction of polysaccharides, simple sugars (fructose and sucrose) and, in particular, aromas. After maceration, the solution was filtered and concentrated to obtain fig syrup (Figure 10).

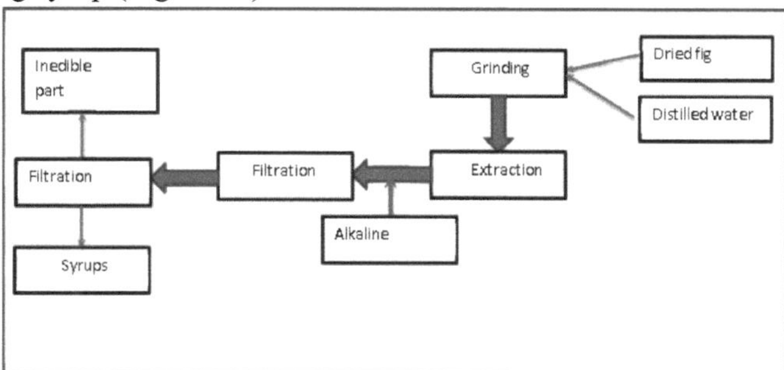

Figure 10: Diagram of fig syrup production (Gabriele et al., 2010).

II.7.2.4. Fermented milk enriched with figs

Today, there is growing interest in the production of fermented milks with therapeutic and nutritional qualities by adding natural additives rich in antioxidant factors and fiber (Simoneliene et al., 2014 and Alaa et al., 2015). According to several authors, consumption of these fortified products can prevent various diseases such as hypertension, hypercholesterolemia, gastrointestinal disorders, diabetes and cancer (Bingham et al., 2003; Ven and Mann, 2004; Pereira et al., 2004).The work carried out by Abd-Eltawab and Ebid, (2019) is part of this approach and aims to develop the production of a brewed probiotic fermented milk enriched with fig (Ficus carica L.). This milk will be a source of dairy ingredients such as milk proteins, calcium, magnesium, vitamin B12 and probiotic bacteria, in addition to the fruit's rich ingredients such as dietary fiber, polyphenols, antioxidants and various minerals. Figur e 11 illustrates the manufacturing flow chart for fig-enriched fermented milk applied in this study (Abd-Eltawab and Ebid, 2019).Addition of fig purées to obtain three different concentrations in addition to the control The mixture is heated to 90° and then cooled in iced water.

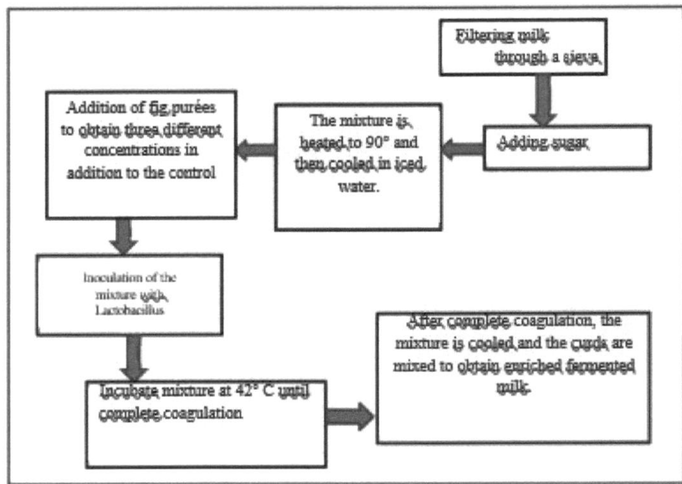

Figure 11: Manufacturing flow chart for fig-enriched fermented milk (Abd-Eltawab and Ebid, 2019).

II.7.2.5. Fig vinegar

Fig vinegar is a traditional fermented product made mainly from fresh or dried figs. Fig vinegar has played an important role in the historical development of vinegar production. There are various recipes and production techniques for fig vinegar.In the study carried out by Sengun (2013), two recipes were developed to assess the microbiological and physicochemical quality of traditionally-produced fig vinegar (Figure12).

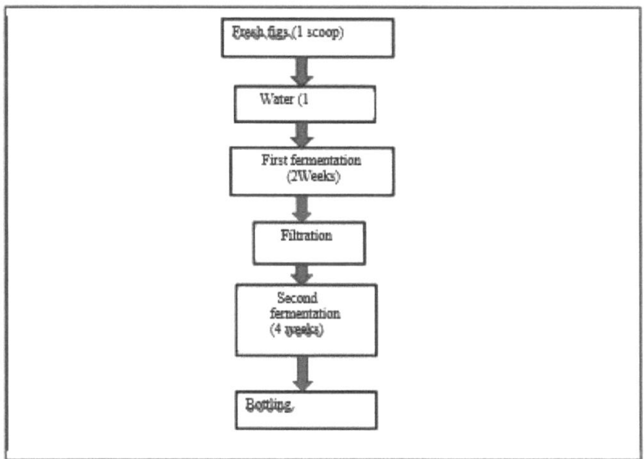

Figure 12: Fig vinegar production diagram (Sengun, 2013).

II.7.2. Fig packaging

Fruit packaging techniques have improved considerably in recent years. As with nectarines, medium-sized cardboard boxes with plastic cells allow each fruit to be individually packaged (Assaf, 2001) (Figure 13). Wrapping and packaging make it possible to :

✓ Contain the product and facilitate its handling and marketing by setting standards for the number or weight of contents in each package.

✓ Protect the product from bruising (impact, compression, friction and damage) and adverse external conditions (temperature, humidity) during transport, storage and marketing.

✓ Provide customers with information on variety, weight, number and approach or quality level of the product, the name of the producer, the country, the region of origin, etc. It is also quite common to include recipes, nutritional values, product barcodes or any other information that can be used to trace the product's origin (Lopez Camelo, 2007).

Figure 13: Packaged product (Ouauiche et chimi, 2005).

Figs for fresh consumption have a limited post-harvest shelf-life of 7 to 14 days under ordinary atmospheric conditions at -1 °C or 0 °C with 95% RH. The post-harvest shelf life of fresh figs is much lower due to the delicate epidermal tissue and the growth of yeasts and molds. It is necessary to use post-harvest techniques that allow the post-harvest shelf life of fresh figs to be extended, such as modified atmosphere packaging (Del Carmen et al ., 2014).

√ **Modified atmosphere packaging**

Modified atmosphere packaging, commonly known as MAP, is used to improve the shelf life of perishable products. In MAP, perishable products are packaged in an atmosphere where the gas composition is different from that of air. As a general rule, a reduction in O2 and an increased concentration of CO2 are used to extend the shelf life of fresh whole and pre-cut products. As a result, MAP effectively extends the post-harvest shelf life of fresh produce by delaying enzymatic browning, reducing respiration rates, minimizing metabolic activity and preserving visual appearance.Substantial shelf-life extension can be achieved by combining MAP with refrigeration. Combined MAP treatment at low temperatures is widely used to maintain the sensory and microbiological quality of fresh produce during long-term storage (Waghmare et al., 2018).

CHAPTER III

PHENOLIC PROFILE OF FIGS AND THEIR ANTIOXIDANT POWER

III.1. Phenolic compounds

Phenolic compounds are common secondary plant metabolites (Caliskan and Polat, 2011). They are known for their protective role against cancer, cardiovascular disease and cataracts (Hollman et al., 1996). These compounds contain an aromatic ring bearing one or more hydroxyl groups (Jean-Jacques Macheix, 1996). Figs are an excellent source of phenolic compounds (59.0 mg EAG/100 g MF) (Marinova et al., 2005). In the study by Slatnar et al. (2011), 8 phenolic compounds were identified in fresh and dried figs. These compounds belong to 4 groups: hydroxy-cinnamic acids, flavan-3-ols, flavonols and anthocyanins. Their levels usually differ not only by variety (56-281.1 mg EAG/100g MS) for dark varieties and (48.6-50 mg EAG/100g MS) for light varieties), but also from one part of the same fruit to another (Solomon et al., 2006). Moreover, Ficus carica is often prepared by peeling (Caliskan and Polat, 2011), while the skin contains a higher quantity of polyphenols (123-463mg EAG/100g MS) (dark varieties) and (41.7-65.5mg EAG/100g) (light varieties) than the flesh. The latter contains between (36.5-100.6mg EAG/100g MS) for dark varieties and (37-59.1mg EAG/100g) for light varieties (Solomon et al., 2006). Phenolic compounds can modulate the activity of a large number of enzymes and cellular receptors (Manach et al., 2004). Their antioxidant activity is mainly due to their redox properties, which enable them to act as reducing agents, hydrogen donors, metal chelators and singlet oxygen scavengers (Rice-Evans et al., 1995).

III.2. Phenolic acids

Phenolic acids are organic compounds with at least one carboxyl function and one phenolic hydroxyl (Bruneton, 1999). They are among the most predominant phenolic compounds in figs, and are concentrated mainly in the skin (Caliskan and Polat, 2011). The two essential groups of phenolic acids are hydroxybenzoic acids and hydroxycinnamic acids, derived respectively from C6-C1 benzoic acid (gallic acid and derivatives) and C6-C3 cinnamic acid (ferulic acid, chlorogenic acid and derivatives) (Figure 14) (Budić- Leto and Lovrić, 2002).

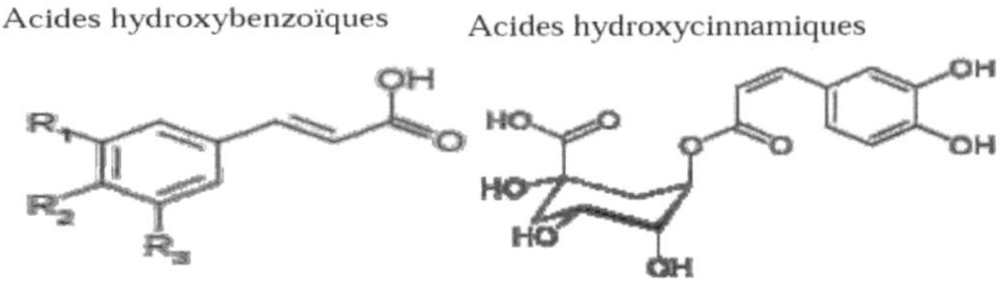

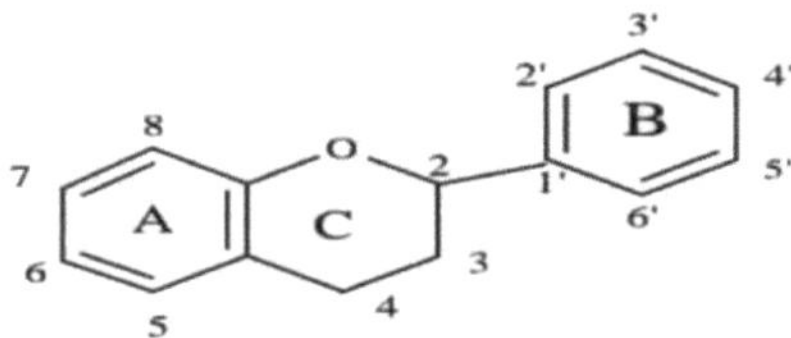

Figure 14: Structure of some phenolic acids found in figs (Handique and Baruah, 2002).

The antioxidant activity of phenolic acids depends on the number and position of hydroxyl groups, and increases with the degree of hydroxylation, as in the case of trihydroxylated gallic acid. However, substitution of hydroxyl groups in positions 3 and 5 by methoxyl groups reduces activity (Tomas-Barberan and Clifford, 2000; Manach et al., 2005; Balasundram et al., 2006).

III.3. flavonoids

The term flavonoid designates a very wide range of natural compounds belonging to the polyphenol family (Seyoum et al., 2006). Structurally, flavonoids can be divided into several classes of molecules, with over 6,400 structures identified (Harborne and Williams, 2000) (Figure 15). They are the most widespread group of secondary metabolites in plants, and consequently one of the most extensively studied (Formica and Regelson, 1995). According to Marinova et al, (2005), the concentration of total flavonoids in fresh figs is 20 mg EC/100g MF . Dried figs are the richest in flavonoids (105.6mg E.Q/100g MF) compared with other dried fruits (Ouchemoukh et al., 2012).

Figure 15: Basic structure of a flavonoid (Coa et al., 1997).

Flavonoids are very powerful antioxidant agents, and have been widely recognized as excellent scavengers of reactive oxygen species from biomolecules such as lipoproteins, proteins and oligonucleic acids (DNA, RNA). This much-studied and widely-acknowledged ability is frequently cited as a key to preventing and/or reducing oxidative stress directly linked to disease.such as cardiovascular disease, carcinogenesis and neurodegenerative disorders. Free radicals are also thought to be involved in the aging process (Quideau et al., 2011). Their antioxidant capacity is enhanced by an increase in the number of hydroxyl groups, O-methylation and a decrease in the number of glycoside groups (Nijveldt et al., 2001; Amiæ et al., 2003). They are generally poorly resorbed in the gastrointestinal tract (Formica and Regelson, 1995).

III.4. The flavonols

Flavonols are the most widely distributed flavonoid compounds, including quercetin, kaempferol, myricetin and apigenin (Figure 16). Dihydroflavonols (dihydrokaempferol, dihydroquercetin) are considered minority flavonoids due to their limited natural distribution (Ghedira, 2005). Quercetin, the most abundant bioflavonoid, is present in good quantities in plant-based foods such as tea, broccoli and red wine (Scalbert and Williamson, 2000). Depending on the source, quercetin can be found as a glucoside (onions), galactoside (apples) or arabinoside (berries) (Erlund, 2004). Of the 20-35 mg of flavonols ingested daily in our diet (Manach et al., 2005), nearly 16 mg are quercetin (expressed as an aglycone) (Hertog et al., 1993). Flavonols are the dominant flavonoids in foods (Manach et al., 2004).

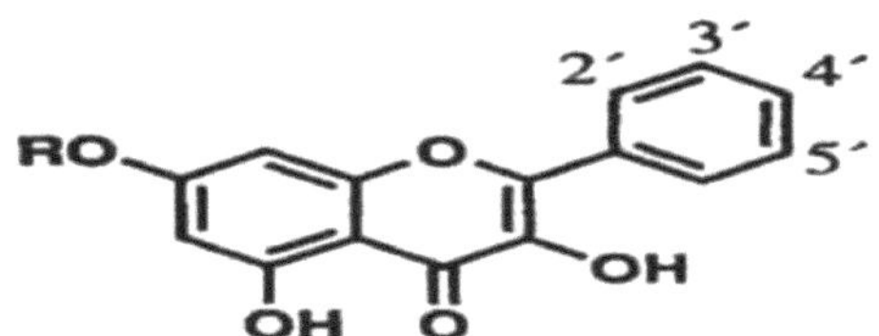

Figure 16: Flavonol structure (Anderson et al., 1996).

According to Del caro and Piga (2008), black fig varieties are richer in flavonols than white fig varieties. HPLC analysis revealed that the majority constituent is rutin (quercetin-3-O-rutinoside). Rutin being a glycosylated form, it is better absorbed by the human body than the non-glycosylated form (quercetin)

(Veberic et al., 2008). Flavonols have beneficial properties in the treatment of heart disease and certain cancers (Vinson et al., 1995), as well as antitumor and chemopreventive activities (Lacaille-Dubois and Wagner, 1996). The antioxidant activity of flavonols is very important (Lugasi et al., 2003), notably in protecting against LDL oxidation, the key treatment for atherosclerosis; they also play an important role in protecting against diabetes in humans, and against oxidative DNA damage (Lean et al., 1999). Quercetin can make a significant contribution to antioxidant potential, as its structure enables stabilization of the aryloxyl radical by donating the hydrogen atom (Zheng and Wang, 2003).

III.5. anthocyanins

The word anthocyanin is derived from two Greek words: anthos, meaning flowers, and kyanos, meaning dark blue. Anthocyanins are distinguished from other flavonoids by their ability to form flavilium cations. Anthocyanins are natural water-soluble pigments and are responsible for the blue, violet, red and orange pigmentation of many fruits and vegetables (Miguel, 2011). Ouchemoukh et al, (2012) have shown in their research that dried figs are the fruit with the highest anthocyanin content (5.9 mg EC/100g DM) compared with other dried fruits such as plums (2 mg/100g DM) and grapes (1mg EC /100g DM). The main anthocyanins in figs are cyanidin 3-O rutinoside and lacyanidin 3-glucoside (Figure17) (Del Caro and Piga, 2008; Dueñas et al., 2008).

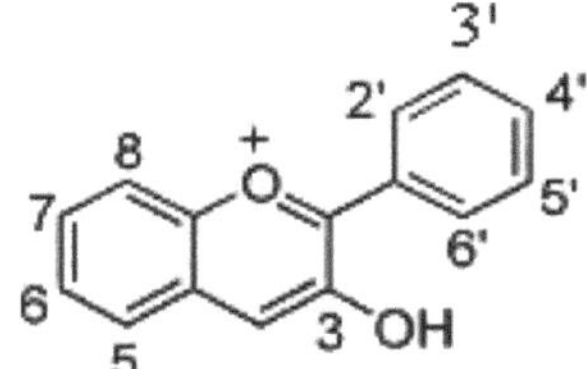

Figure 17: Structure of anthocyanidins (Robards, 2003).

The basic structure of anthocyanins is characterized by a flavone ring generally glucosylated in the C-3 position (Vandi et al., 2016). The antioxidant effect of anthocyanins is explained in part by free radical scavenging and metal chelation. In addition, cyanidol forms a pigmentation complex with DNA, protecting this critical molecule from oxidative damage. Anthocyanins inhibit the proteolytic enzymes that degrade collagen (elastase, collagenase), which explains their vasoprotective and anti-oedematous properties. They are also vein-active compounds with a vitamin "P" property. ". Anthocyanins block the production of

NO (nitric oxide) from polynuclear cells.neutrophils during the early phase of inflammation and are considered anti-inflammatory molecules (Derbel and Ghedira, 2005).

III.6. The tannins

Tannins are water-soluble polyphenols with a molar mass of between 500 and 3000 g/mol. In addition to the classic reactions of phenols, tannins have the property of precipitating alkaloids, gelatin and other proteins. They are found in all parts of the plant, bark, wood, leaves, fruit and roots in varying concentrations. (Vandi et al., 2016). According to Mahmoudi et al., (2018) the condensed tannin contents of nine Algerian fig varieties studied were 0.388 µg of catechol equivalent / g PF for the skin and 18.468 µg EC/g PF for the pulp. Structurally, tannins are divided into two groups:

III.6.1. Condensed tannins (proanthocyani- dine)

Are polymers of flavanic units most often linked together by c4-c8 bonds. The precursors are flavan -3ols (catechin and epicatechin) and flavan -3.4 diols (Figure18). This class of tannins is the most common in the plant world (Zimmer and Cordesse, 1996).

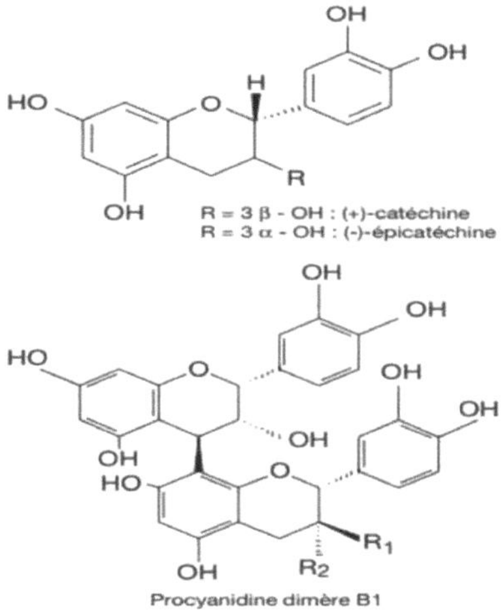

Figure 18: Structure of catechin, epicatechin and condensed tannin (proanthocyanidol) (Derbel and Ghedira, 2005).

III.6.2. hydrolysable tannins

These are gallic acid esters and hexahydroxydiphenic acid esters. By hydrolysis (acid, alkaline or enzymatic), the phenolic acids released are gallic acid or ellagic acid, from gallic tannins (gallo-tannins) and ellagic tannins (ellagitanins) respectively (Figure19) (Zimmer and Cordesse, 1996).

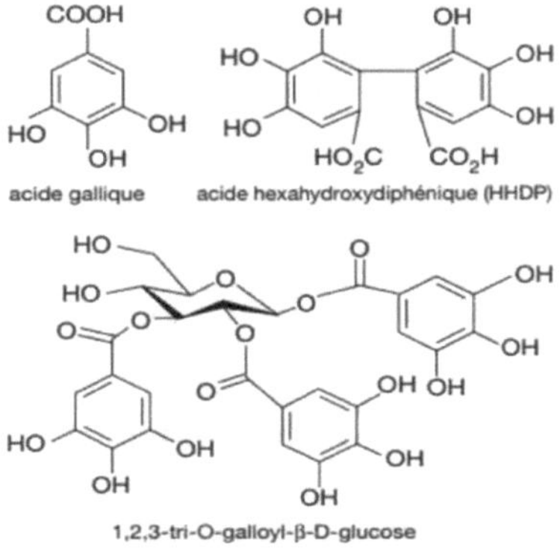

Figure 19: Structure of gallinic acid, hexahydroxydiphenic acid and a tannin gallic (Derbel and Ghedira, 2005).

Tannins have significant antioxidant power. Hydrolyzable tannins inhibit lipid peroxidation, while condensed tannins inhibit superoxide formation. Tea catechols trap free radicals, chelate metal ions and preserve other antioxidants such as vitamin E. These polyphenolic compounds inhibit the enzymatic activities of protein kinase C, 5-lipoxygenase and angiotensin-converting enzyme, and are active on thermogenesis. Some tannins also have vitamin "P" (or quercetin) properties, the former name of vitamin C2 (Derbel and Ghedira, 2005).

III.7. carotenoids

Carotenoids are natural pigments belonging to the tetra-terpene family. They consist of four terpene units, each containing 10 carbon atoms. Their structure contains double bonds and sometimes ring(s), with or without attached oxygen atoms (Arvaniti et al., 2019). They are responsible for the red, orange and yellow coloring of fruits and vegetables (Rao and Rao, 2007). There are two main classes of carotenoids:

• Xanthophylls contain one or more oxygen atoms. Lutein is the most popular compound in this class.

• Oxygen-free carotenes consist solely of carbon and hydrogen atoms.β-carotene
is the main compound in the carotene group (Arvaniti et al., 2019).

Carotenoids are known to have a significant impact on health by acting as
lipophilic antioxidants. They are effective deactivators of electronically excited
sensitized molecules involved in the generation of oxygen singlets and the
peroxyl radical (Young et al., 2001; Tapiero et al., 2004).

The antioxidant properties of carotenoids are associated with their radical
scavenging properties and exceptional singlet oxygen deactivation capabilities.
Carotenoids can scavenge free radicals through three reactions: electron transfer,
hydrogen abstraction and addition (El-Agamey et al., 2004).

$$CAR + ROO°CAR°+ + ROO^-$$

$$CAR + ROO°CAR° + ROOH$$

$$CAR + ROO°ROOCAR°$$

Some of the carotenoids have provitamin A properties to varying degrees, such
as the carotenes whose most active form is β-carotene (trans form), as well as
cryptoxanthin, and their derivatives (Couplan, 1998). Fresh and dried figs
contain a variety of carotenoid compounds (figure20) (Arvaniti et al., 2019) .
According to the study by Kakhniashvili et al, (1987) on the carotenoid profile
of figs, the results showed that figs contain carotenoids between 1.59 and 4.32
mg/100 g, of which the two major ones in figs are lutein and α-carotene
(figure21).

Lycopene

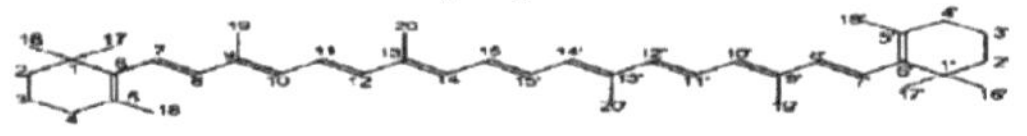

β-Carotene

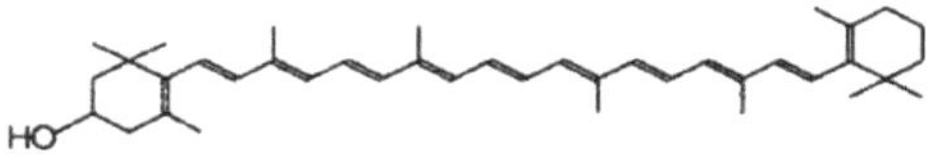

β-Cryptoxanthin

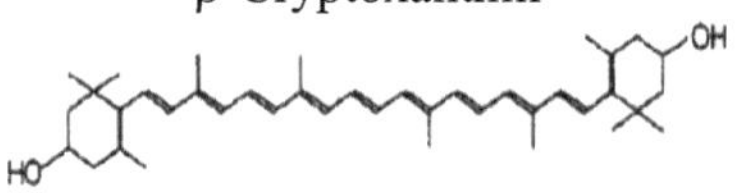

Zeaxanthin

Figure 20: Structure of some carotenoids (Arvaniti et al., 2019).

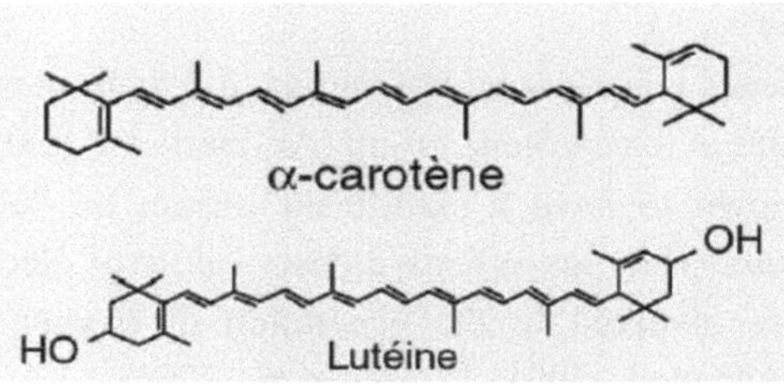

Figure 21: Structures of α-carotene and lutein (Borel et al., 2005).

CHAPTER IV
PREVIOUS STUDIES ON FICUS CARICA

Edible figs, Ficus carica L., play an important role in human and animal nutrition worldwide. Humans have consumed the fruit of these trees since the earliest times, and have used it and other parts of the tree for medicinal purposes. Recently, much work has focused on studying the quality of this fruit, whether nutritional, physicochemical or pharmacological.

IV.1. Previous studies on nutritional quality

As a seasonal food, the fig represents an important constituent of the Mediterranean diet and has a very diverse composition. It contains a low amount of lipids and is an excellent source of minerals, vitamins and dietary fiber (Mahmoudi et al., 2018).It is well known that the nutritional composition of foods is of direct interest to the consumer. For this reason, the analysis of nutritional values makes it possible to establish or confirm the nutritional content of food products. These analyses also make it possible to add value to the composition of our foods, by highlighting their nutritional qualities (fiber content, specific minerals, vitamins, sugars, etc.). With this in mind, a number of studies have focused on the nutritional characterization of Ficus carica.Food proteins of plant origin play a major role in the human diet, even if their biological value is lower than that of animal proteins. They fulfil essential functions as enzymes, nutrient transporters and even defence agents, as well as being the cause of some allergic reactions (Lacroix, 2008). Several authors (Abou-Farrag et al., 2013 and Pereira et al., 2017) have assayed fig proteins in particular.These authors generally rely on the method of Bradford (1976). The principle of the assay is based on the binding of Coomassie Blue G-250 to aromatic amino acid residues present in proteins to give a colored complex which absorbs ultraviolet rays at 595nm. The subsequent soluble protein content is determined by reference to a calibration curve obtained from a solution of bovine serum albumin (BSA).The study conducted by Abou-Farrag et al, (2013) on three varieties of Egyptian fresh figs showed that protein content varied from 4.15% to 5.38%. These results are slightly lower than those obtained by Pereira et al. (2017), who reported mean values of ranging from 4.4% to 6.7% for nine fresh fig varieties grown in Extremadura (Spain). Fruits contain 0.1 to 1.5% nitrogen compounds, 35 to 75% of which are protein. Free amino acids account for an average of 50% of soluble nitrogen compounds. Amino acid composition is typical for each fruit, and can therefore be used for the analytical characterization of fruit products (Belitz et al., 2009).Colorimetric determination of amino acid content is based on the reaction with ninhydrin. The intensity of the color formed is proportional to the quantity of amino acids in the extract (Yemm and Cocking, 1955).Amino acid content is then determined by reference to a calibration curve obtained from a standard solution (glycine solution, for example). Results are expressed in g per 100 g MS.Determination of total amino acid content in the study by Huang et al. (2010), which analyzed the nutritional quality and antioxidant activity of 13 edible and medicinal plants from Hong Kong, including fresh figs, revealed a content of 9.29 g/100 g. An average lower free amino acid content of fresh figs of 0.82 g/100 g was reported by Favier et al. (1993).Total carbohydrates include all polysaccharides, oligosaccharides and monosaccharides. They make up the major part of our diet and are supplied mainly by fruit (Lee et al., 1970). Fruit carbohydrate concentration is of great interest, because of its influence on organoleptic properties and is a criterion for assessing ripeness; it also conditions fruit stability and storability (Golubev et al., 1987; Jiang et al., 2013).Total sugars are generally determined by the phenol-sulfuric acid method of Dubois et al. (1956). When heated in the presence of strong acid, carbohydrates dehydrate to form furan derivatives (furfural in the case of pentoses and hydroxymethylfurfural in the case of hexoses), which condense with phenol to give a yellow-orange complex with an absorption maximum between 480 and 490 nm. Total carbohydrate content is calculated by reference to a calibration curve previously established with a standard (glucose). Results are expressed in g standard equivalent (glucose) per 100 g MS.The results of the study carried out by Trad et al, (2014) on five Tunisian fig cultivars reported total sugar contents ranging from 11.8 to 16.4 g/100 g. In a study conducted by Mahmoudi et al, (2018) on nine varieties of Algerian fresh figs, total sugar levels in the pulp were higher (9.453 g / 100 g to 26.016 g / 100 g) than those obtained in the skin (0.958 g / 100 g to 11.594 g / 100 g).The quantity of sugars in figs is particularly important for the food industry: for the production of jam and dried figs (Darjazi, 2011). The study carried out by Aljane et al, (2006) on fourteen fresh fig varieties from southern Tunisia, showed that the values obtained for glucose varied from 1.216 to 6.133g / 100 g fresh matter and the values obtained for fructose varied from 1.916 to 4.658 g / 100 g fresh matter. On the other hand, the study by Awatef et al (2010), carried out on three fresh fig cultivars from south-eastern Tunisia, showed a glucose content ranging from 8.53 to 10.46g / 100 g fresh matter and a fructose content ranging from 7.25 to 9.35g / 100 g fresh matter. Minerals are micronutrients involved in many biochemical processes, and an appropriate intake of these minerals is essential for the prevention of diseases related to mineral insufficiency (Mahmoudi et al., (2018). Fig is considered a good source of minerals especially potassium and calcium (Lim, 2012).Determination of mineral content is based on mineralization, i.e. the recovery of ash, obtained by incinerating fig samples at 550°C in liquid form. Determination of mineral content is based on the standard (NF V 05-113, 1972), using atomic absorption spectrophotometry.Several authors have measured mineral salts in figs, including Aljane et al. (2006), who found that the mineral salt contents of fourteen varieties of fresh figs harvested in southern Tunisia ranged from 541.27 to 875 for potassium (K), from 151.76 to 448.07 for calcium (Ca), from 51.69 to 91.22 for magnesium (Mg), from 19.7 to 57.52 for sodium (Na), and from 0.77 to 2.08 for zinc (Zn). These values are expressed in mg/100 g dry matter. On the other hand, Aksoy et al (1987) and Ozer and Derici (1998) found mineral salt contents for Turkish fig cultivars ranging from 680 to 1050 for potassium (K), 167 to 333 for calcium (Ca), 11 to 107 for magnesium (Mg), 20 to 67 for sodium (Na) and 0.8 to 2.0 for zinc (Zn). These values are expressed in mg / 100 g dry matter.Figs are rich in water-soluble vitamins B1, B2 and C and fat-soluble vitamins such as vitamin A (Farahnaky et al., 2009), vitamin D2, vitamin D3, vitamin K1 and α- Tocopherol as well as γ-Tocopherol, δ-Tocopherol and α-Tocopherol acetate. The levels of these vitamins vary between different parts of the fruit. Vitamin C works with vitamin E and the enzyme glutathione peroxidase to scavenge radicals (Cheikh-Traoré, 2006). It reacts not only with hydroxyl radicals, but also with superoxide radicals and their protonated form, and scavenges peroxyl radicals (Gardès-Albert et al., 2003). During these reactions, ascorbate is oxidized to the ascorbyl radical, which is then recycled, at least in part, by dismutation (Gardès-Albert et al., 2003).Ascorbic acid is the reduced form of vitamin C. It has two isomers: L-ascorbic acid and D-ascorbic acid. Only the L form is metabolized efficiently in humans (Sekli-Belaidi, 2011). It has several physiological roles and is involved in the elimination of carcinogens and dangerous nitrosamines (Liao and Seib, 1988).its determination is based on an acid redox reaction using 2,6-dichloro-indophenol (DCIP). In the presence of ascorbic acid, DCIP (blue) is reduced to DCIPH2 (colorless). Excess DCIP gives a characteristic pink coloration with an absorption maximum at 515 nm (Tabart et al., 2010).The study conducted by Mahmoudi et al, (2018), on nine varieties of Algerian fresh figs showed that vitamin C concentrations in pulps and peels varied from 1.33 to 10.67 mg/100 g. In contrast, a study by

Abou-Farrag et al (2013) on three varieties of fresh Egyptian figs showed vitamin C concentrations ranging from 2.16 to 6.71 mg/100g. Dietary fibers include lignin and carbohydrates such as cellulose, hemicelluloses, pectins, resistant starches and non-digestible oligosaccharides. Figs contain both types of fiber (soluble and insoluble). It is an excellent remedy for constipation (Bidri, 2018). Its fibers are highly effective in stimulating the intestines (Du Toit et al., 2001).According to the method described by AOAC 985.29 and adopted by Pereira et al., (2017), the principle of fiber assay consists in a first step of sample digestion using an enzymatic cocktail (α amylase, protease and amyloglucosidase), followed by treatment with ethanol to precipitate soluble fibers and extract proteins and glucose. The resulting precipitate is washed, dried and weighed.The results of Pereira et al, (2017), showed that the dietary fiber content of nine fresh fig varieties grown in Extremadura (Spain) varied from 4% to 7.4%. These results are superior to those obtained by Abou-Farrag et al, (2013) who reported fiber contents ranging from 3.91% to 5.17%.

IV.2. Previous studies on physicochemical quality

The main function of a fig tree is to produce delicious, nutritious fruit, either fresh or dried. During the harvest period, fresh produce is plentiful, but the rest of the time it's hard to find. Most fruits, such as figs, remain consumable for only a short time if they are not stored quickly (James and Kuipers , 2003). The development potential of increased fig production is linked to the introduction of modern agricultural techniques and the selection of productive cultivars with determination of physico-chemical parameters linked to fruit quality (Pereira et al., 2017). These parameters are used as indicators of ripeness threshold (Chahidi et al., 2008), such as pH, moisture, titratable acidity, total soluble solids (TSS), ripening index (RI), weight, ash and firmness.pH is an important parameter in food quality control. It is a criterion for classifying fruit and vegetables, and plays a limiting role in their preservation (Board, 1987). According to Abbas et al,(2016), the pH values recorded were between 4.0 and 4.7 for figs from Pakistan. The values obtained in this study were lower than those reported by Caliskan and Polat (2008), who obtained values between 4.6 and 5.4 for figs from Turkey. However, it is interesting to note that Simsek (2009), who measured the pH of the juice of seven varieties of fresh figs from southern Turkey, harvested in two successive years, reported values ranging from 4.67 to 6.04.In another investigation, Abou-Farrag et al ., (2013) obtained pH values for three fresh fig varieties from Egypt ranging from 5.28 to 5.50.

Slightly higher values ranging from 5.16 to 6.39 were recorded by Pereira et al, (2020) for nine fig varieties from Spain. This discrepancy can be explained by the influence of environmental conditions on quality.fruit (Valero and Serrano, 2010).In addition to pH, humidity is another important parameter in the enhancement of physicochemical quality. Fresh figs are highly susceptible to microbial spoilage, even under cold storage conditions, and are highly perishable at room temperature, due to high humidity (Abou-Farrag et al., 2013). Moisture represents the difference between the weight of the sample before and after drying.when weight is constant (Al Askari et al., 2012). Indeed, to determine the water content, the fresh material is dried in an oven at a temperature of 103±2C°.isotherm ventilated at atmospheric pressure to a virtually constant measurement. The study carried out by Aljane and Ferchichi (2009), on a dozen varieties of Tunisian fresh figs showed moisture contents ranging from 63.46 to 83.47%. The values obtained in this study were lower than those obtained by Abbas et al, (2016), whose moisture content ranged from 78.21 to 85.94% for four fig varieties from Pakistan . Values close to those obtained by Abbas et al, (2016), were reported by Abou-Farrag et al, (2013) with moisture contents between 77.81 and 82.91% for three fresh fig varieties from Egypt.In physicochemical characterization, it is also important to measure titratable acidity. It is an important element in determining the harvest date. It is an indicator of fruit maturity in relation to sugars. In general, organic acid content decreases during ripening (Chahidi et al., 2008). To determine titratable acidity, the food is first diluted if the concentration of organic acids is high. A portion of the diluted solution is then titrated with a standardized 0.1N NaOH solution, in the presence of an indicator. For over-colored foods, a pH meter is used to detect the turning point (pH 8.3 for phenolphthalein).The study by Pereira et al, (2017) showed that titratable acidity data at various stages of ripening for nine fig varieties grown in Extremadura in western Spain ranged from 0.1 to 0.2 g/100 g . The values obtained in this study were lower than those obtained by Caliskan and Polat (2012), who reported values ranging from 0.11 to 0.35 g/100 g for figs from the eastern Mediterranean region of Turkey. Simsek (2009) reported values between 0.14 and 0.29 g/100 g for seven varieties of fresh fig from southern Turkey.According to Ercisli et al. (2012), the titratable acidity of fig genotypes and cultivars from northeastern Turkey varies between 0.16 and 0.47%. It should be noted that, to obtain better quality figs, titratable acidity should be between 0.22 and 0.30% (Ercisli et al., 2012). Another parameter, total soluble solids (TSS) are among the main physico-chemical properties to be analyzed also in fresh figs. An increase in TSS content during storage may be due to the conversion of polysaccharides to soluble sugars (Abbas et al., 2016). Total soluble solids (TSS) are generally measured using a refractometer. According to Abbas et al.,(2016), the total soluble solids (TSS) of four fig varieties from Pakistan varied from 9.65 BRIX° for the English variety, 10.63 BRIX° for the dark brown variety, 12.82 for the dark black variety and 10.70 BRIX° for the wild variety. It was observed that total soluble solids in all samples increased during 30-day storage. Total soluble solids in the English variety increased from (9.4 to 10.2 BRIX°), for the dark brown fig variety from (10.2 to 11.10 BRIX°), for the dark black variety from (12.2 to 13.6) and for the wild cultivar from (10.2 to 11.14 BRIX°). These results were correlated with those of Del Carmen villalobos et al, (2016), who also found an increase in SST values during the ripening process for the two fig cultivars studied"Cuello Dama Blanco and Cuello Dama Negro. It is during ripening that the organoleptic quality of the fruit develops (accumulation of sugars and acids, production of aromas, changes in texture, etc.). For this reason, assessment of the ripening index, which represents the SST/AT ratio in the physico-chemical analysis of fresh figs, is important, as this ratio is an indicator of consumer acceptability and fruit quality, since the perceived sweetness of ripe fruit depends on the SST/AT ratio (Pereira et al ., 2017) .

According to Pereira et al. (2017), MI values for nine fresh fig varieties from Spain ranged from 108 to 221. The SST / AT ratios obtained in this study were much higher than those obtained by Abou-Farrag et al., (2013) for three fig varieties from Egypt who reported values ranging from 90.6 to 155.8.Weight is another physical parameter not to be overlooked. Indeed, from a commercial and economic point of view, a large fruit is synonymous with good production and is more easily sold (consumer attraction). Smaller fig fruits are more suitable for jam-making, while larger fruits are more suitable for fresh consumption.When analyzing seven varieties of fresh figs from southern Turkey, Simsek (2009) reported weights ranging from 30.88 to 56.29 g. Pereira et al, (2017) obtained a slightly wider weight range for fresh figs from Spain (36.9 to 56.8 g) . Furthermore Crisosto et al.,(2010) reported weights ranging from 35.60 to 55.60 g for figs from the USA, while for Caliskan and Polat (2012), weights ranged from 12.30 to 99.40 g. The differences between the results can be explained by varietal effect, fig caprification, geographical and ecological variations including the effect of growing conditions. Ash determination is also an important parameter. Total ash is the residue of mineral compounds remaining after incineration of a sample containing organic substances of animal, vegetable or synthetic origin. Ash represents approximately 1 to 5% of the mass of a food on a wet basis. The study carried out by Abou-Farrag et al, (2013) on three Egyptian fig varieties showed ash contents ranging from 2.53 to 3.08%. The values obtained in this study were much higher than those reported by Guvanc et al, (2009) who obtained values ranging from 0.48 to 0.85% for fresh Turkish figs. It should also be noted that several studies have shown that ash contents increase slightly during storage (Abbas et al.,(2016), Sandhu et al.,(2001)).Measuring fruit firmness is also an important parameter. It is one of the criteria for estimating the best time to harvest. It gives a quantified indication of a fruit's hardness or tenderness, and consequently establishes the optimum ripening point to avoid fruit becoming more susceptible to damage during transport and storage. Indeed, fruit ripening, like certain alterations due to deficiencies, disease or unfavorable storage conditions, results in changes in the composition of inter-membrane lamellae, the membranes themselves or cell turgor, which can ultimately be detected by changes in tissue crush resistance (Pereira et al ., 2017).According to Pereira et al, (2017), the average firmness values of nine fig varieties in Extremadura in western Spain ranged from 1.9 to 7.1 Nmm -1. The variety"Brown Turkey" showed the highest firmness value (7.1 Nmm -1). Conversely, "Cuello Dama Blanco" was the variety with the lowest firmness value (1.9 Nmm-1). Significant differences were also observed between ripening stages. In general, firmness was strongly affected by fruit ripening stage, decreasing progressively with maturity. These results are in agreement with those of Del Carmen Villalobos et al, (2016), who also found a decrease in firmness throughout the storage period for the two fig cultivars studied "Cuello Dama Blanco" and "Cuello Dama Negro" .

IV.3. Previous studies on phenolic compounds and antioxidant activity

Plants are known to possess so-called "secondary" metabolites, as opposed to primary metabolites (proteins, carbohydrates and lipids). These metabolites secondary metabolites are classified into several major groups, including phenolic compounds. A number of studies have focused on the extraction of these secondary metabolites from figs and the valorization of these compounds by evaluating different activities, notably antioxidant activity. The extraction of phenolic compounds is a crucial step in the valorization of active ingredients. It is influenced by their chemical structure, the extraction method, the particle size of the sample, storage time and conditions, and the presence of interferents (Naczk and Shahidi, 2004). It also depends on the extraction solvent, which is also one of the parameters that can affect polyphenol extraction (Troszynska et al., 2002). Extraction can be carried out using a number of solvents, including water, methanol, ethanol and acetone. Aqueous solvents give better extraction yields than absolute solvents (Spignon et al., 2007). It has also been shown that of the three solvents, chloroform, ethyl acetate and methanol, methanol is the most widely used for the extraction of antioxidant compounds (Osawa and Namiki, 1981). Indeed several authors have used methanol for the preparation of fig extracts (Mahmoudi et al., (2018), Solomon et al., (2006), Jokić et al., (2014)

The subsequent content of total phenolic compounds in fig extracts is generally estimated by the Folin-Ciocalteau method. This spectrophotometric method is one of the methods adopted in most works concerning the quantification of total polyphenols (Mahmoudi et al., (2018), Caliskan and Polat (2011),Meziant et al., (2015), Spigno et al., (2007). This test, based on a redox reaction, can also be considered as a method for assessing antioxidant activity (Prior et al., 2005). As a result, extracts richer in phenolic compounds can also be considered the most antioxidant. In fact, during the assay, the polyphenol content is estimated from the regression equation of the calibration range established with a standard such as gallic acid, and the content is then expressed in milligrams of standard equivalent (e.g. gallic acid) per gram of crude extract or per gram of fresh fig weight, or even per gram of dry weight.In Algeria, according to previous investigations, the average level of polyphenols contained in extracts derived from ficus carica generally varied between 217 and 342 mg EAG / 100 g dry weight (Meziant et al., 2015). In Tunisia, Aljane (2018) and in Turkey, Caliskan and Polat (2011) detected significantly lower levels, ranging from 51.50 to 100.23 mg and 28.6 to 211.9 mg EAG / 100 g fresh matter respectively.Other studies carried out on Ficus carica polyphenols have revealed the presence of the following phenolic compounds: gallic acid, chlorogenic acid, syringic acid, (+)-catechin, epicatechin and rutin, with the highest rutin content (up to 28.7 mg per 100 g PS) (Veberic et al., 2008).Flavonoids are the most abundant and well-studied class of polyphenols. Most of the research undertaken on fig extracts has also attempted to estimate the content of these metabolites. Quantitative estimation of total flavonoids contained in extracts is generally carried out using the aluminum trichloride method ($AlCl_3$) (Mahmoudi et al.,(2016), Harzallah et al., (2016), Meziant et al., (2015), Jokić et al., (2014)).The quantity of flavonoids contained in the extracts is calculated using a calibration range established with a standard. Results are expressed in standard milligram equivalents per gram of crude extract or per gram of fresh fig weight, or even per gram of dry weight.Average flavonoid contents in Algerian fig extracts from the Bejaia region ranged from 11.13 to 19.20 mg Quercetin Equivalent (QE) / 100 g dry weight (DW) (Meziant et al., 2015). The work of Mahmoudi et al, (2018) indicated average flavonoid contents in fig extracts from the Bouira region of between 11.56 and 16.36 mg EQ / 100 g PS . Two fig varieties grown in the Tirana region of Albania had flavonoid contents ranging from 42.47 to 269.54 mg catechin equivalent (CE)/100 g fresh weight (FP) (Hoxha and Kongoli, 2016). In another investigation, Solomon et al, (2006), reported total flavonoid values for six fresh fig varieties ranging from 2.1 to 21.5 mg EC/100 g FP. On the other hand, Petkova et al, (2019) showed that flavonoid levels in the fresh Bulgarian fig sample were higher (11.4 mg EQ/100 g PF) compared with frozen fig samples (3.6 mg EQ/100 g PF) and fig jam (6.6 mg EQ/100 g PF).The study carried out by Aljane and Sdiri (2014), on a dozen varieties of fresh figs grown in south-eastern Tunisia showed flavonoid contents ranging from 4.30 to 9.60 mg catechin / 100g PF. Anthocyanins also belong to the class of phenolic compounds, and are responsible for the orange, pink, red, violet and blue color of many fruits and vegetables (Sass-Kiss et al., 2005; Rogez et al., 2011). These pigments are soluble in polar solvents (Kong et al., 2003).It should be noted that the most common procedure for extracting anthocyanins uses solvents such as methanol or ethanol with varying proportions of hydrochloric acid (HCl 0.1-1%). The use of an organic solvent has the advantage of inactivating enzymes present in plant tissues and facilitating subsequent processing of the extract obtained, as it is easily evaporated. Nevertheless, depending on the nature of the sample, it may be necessary to add water in order to achieve complete anthocyanin extraction.According to Mahmoudi et al. (2018), the anthocyanin content of nine Algerian fig varieties studied ranged from 25.4 to 54.78μg of cyanidin-3 rutinoside equivalents. / g of fresh weight for the skin and between 4.488 and 8.077 μg of cyanidin-3 rutinoside equivalents / g of fresh weight for the pulp. These levels are lower than those recorded by Duenas et al, (2008), who studied five different varieties of fig in Spain and whose total anthocyanin content in the skin varied between 32 and 97 μg of cyanidin-3 rutinoside equivalents / g and between 1.5 and 15 / μg of cyanidin-3 rutinoside equivalents/ g in the pulp.Another study by Aljane and Sdiri (2014), on ten varieties grown in south-eastern Tunisia showed anthocyanin contents ranging from 0.55 to 9.16 mg cyanidin -3-glucosie equivalents (ECG) / 100g PF. The study carried out by Caliskan and Polat (2011) on several fig varieties including San Pedro, Smyrna type from Hatay province located in the eastern Mediterranean region of Turkey showed anthocyanin contents ranging from 0.7 to 298.9 mg cyanidin-3 rutinoside equivalent / g PF. Also, according to phytochemical investigations on figs, anthocyanidins including cyanidin 3-O-rutinoside and cyanidin 3-O-glucoside were found. Moreover, according to a study by Del Caro and Piga (2008), cyanidin 3-O-rutinoside was the second most representative compound in figs after rutin. Duenas et al (2008) used HPLC-DAD-MS to detect other types of anthocyanins in addition to the anthocyanins already mentioned: pelargonidin 3-glucoside, pelargonidin 3-rutinoside and petunidin 3-rutinoside. Black fig peels showed appreciable cyanidin 3-O-rutinoside content, while cyanidin 3-O-glucoside was detected only in fig peels. These data confirm those reported by Solomon (Solomon et al., 2006).Condensed tannins, also known as proanthocyanidins - polymeric derivatives of flavan-3-ol - are also estimated in fig extracts. The principle of condensed tannin determination involves depolymerizing tannins in an acidic methanolic medium, and after reaction with vanillin, converting them into red anthocyanidols that can be easily analyzed at 500 nm Sun et al. (1998). Concentrations of condensed tannins are deduced from a calibration range established with the standard (e.g. catechin) and are expressed in milligrams of standard equivalent per gram of dry plant matter (mg EC/g MS) or even per gram of fresh fig weight.According to Mahmoudi et al, (2018), the condensed tannin contents of nine Algerian fig varieties studied were 0.388 µg of catechol equivalent / g PF for the skin and 18.468 µg EC/g PF for the pulp. Harzallah et al, (2016) working on three extracts namely peel juice, pulp and whole fruit of three fig varieties (the Kohli, Hamri and Bidhi variety) grown in Tunisia obtained condensed tannin contents that varied according to the part studied. Indeed, the authors reported that the highest contents were obtained with the Bidhi variety, with 75.98 mg EAT / g PF in the peel juice, 65.87 mg EAT / g PF in the pulp and 60.35 mg EAT / g PF in the whole fruit. Carotenoids are fat-soluble pigments (Ferreiro-Vera et al., 2011), synthesized by plants but not by the human body. They are found in small quantities in fruit and vegetables, and are responsible for the color yellow, orange and red. Carotenoid determination is performed after an extraction step using appropriate solvents, and the concentration is subsequently estimated by reference to the calibration curve obtained using a standard such as β-carotene.Ficus carica represents one of the sources of carotenoids with around 11 mg EβC/100g MS according to Ouchemoukh et al., (2012). Levels of 3.52 µg /g PF, 1.73 µg /g PF and 1.94 µg /g PF were recorded for fresh, frozen figs and fig jams from Bulgaria respectively according to Petkova et al., (2019). Furthermore, the study carried out by Aljane and Sdiri (2014), on a dozen varieties of fresh figs grown in south-eastern Tunisia showed carotenoid contents ranging from 0.04 to 0.56 µg / 100g PF.In another study, the following carotenoids were highlighted: cryptoxanthin (with a content of 3.01 µg /100g PF), β-carotene (with a content of 7.19 µg /100g PF) and α-carotene (with a content of 78.24 µg /100g PF) Ouchemoukh et al.,(2016). Solomon et al. (2006) reported the presence of several carotenoids, including lutein, cryphoxantin and lycopene. The difference in phenolic compound content noted between the different studies may be due to several factors, including extraction and analysis methods, geographical origin (sunshine levels), degree of ripeness and storage conditions (Rodriguez Montealegre et al., 2006).Many diseases such as cancer, inflammation, atherosclerosis, Alzheimer's and Parkinson's are linked to ROS-mediated damage to biological macromolecules, resulting from an imbalance between radical-generating and radical-scavenging systems (Taubert et al., 2003).

To this end, a number of studies are also assessing the antioxidant potential of polyphenolic extracts. Antioxidant activity can be assessed by a number of different methods, including 2, 2-azinobis (3-ethylbenzothiazoline) - 6-sulfonic acid (ABTS), 1,1-diphenyl- 2-picrylhydrazil (DPPH) and the iron reducing power test (Aljane,2018).The DPPH test is widely used to determine the anti-free radical capacity of various samples. In fact, when a DPPH solution is mixed with a substance that can donate a hydrogen atom, this gives rise to the reduced form with a loss of violet color. The discoloration will be directly proportional to the number of protons captured, and can be monitored by reading the absorbance of the reaction medium at 517 nm. This makes it possible to assess the rate of DPPH reduction and thus provides a practical means of measuring the antioxidant power of the extracts under study.The study carried out by Arumugam et al, (2018) on skins and pulps showed that the DPPH radical scavenging activity of the skin extract was higher (72.84%) compared to the pulp extract (38.51%).According to Lahmadi et al, (2019), the anti-free radical activity profiles obtained revealed that Ficus carica extract possesses concentration-dependent anti-free radical activity. The percentage of inhibition ranged from 11.31% to 87.03%. The study conducted by Aljane, (2018) working on several Tunisian fig varieties reported inhibition percentages ranging from 11.36% for the Besbessi variety to 64.737% for the Bouharrag variety with an overall average for all accessions of 36.62%.Like the DPPH test, the determination of iron's reducing power is a simple analysis in its application and is in every way complementary to the DPPH test. This method is based on the ability of extracts to reduce ferric iron (Fe^{3+}) to ferrous iron (Fe^{2+}). The mechanism is known to be an indicator of electron-donating activity, characteristic of the antioxidant activity of polyphenols (Yildrim et al., 2001).

Indeed, the ferric-tripyridyltriazine complex is reduced to the ferrous-tripyridyltriazine form in the presence of antioxidants; the complex loses its yellow color for a dark blue. This coloration, measured at 700 nm, is proportional to the concentration of antioxidants present in the samples.Results obtained by Lahmadi et al,(2019), on Ficus carica extract from Morocco showed that iron reduction was proportional to the concentrations used. Absorbances ranged from 0.113 to 0.494. The same finding was reported by Harzallah et al, (2016) on different varieties of figs grown in Tunisia, with absorbances varying according to the part of the fruit from 2.33 (whole fruit) to 3.36 (skin). The ABTS-$^+$ radical-cation trapping method is based on the compounds' ability to trap the ABTS$^+$ radical-cation, (ammonium salt of 2,2'-azinobis(3-ethylbenzothiazoline)-6-sulfonic acid), which exhibits an absorption spectrum in the visible range with three maxima at 645, 734 and 815 nm (Thaipong et al., 2006). When reacted with potassium persulfate (K2S2O8), ABTS forms the blue to green radical ABTS-$^+$. The addition of an antioxidant will reduce this radical and cause the mixture to discolor. The decoloration of the radical, measured spectrophotometrically at 734 nm, is proportional to the antioxidant concentration. The cationic radical ABTS-$^+$ is formed by the removal of an electron (e-) from a nitrogen atom of ABTS. In the presence of an H- donating antioxidant, the nitrogen atom concerned traps an H-, leading to ABTS$^+$, which causes the solution to discolor (Lien et al., 1999). the study carried out by Arumugam et al, (2018) on ficus carica skins and pulp showed that the percentage of ABTS radical-cation scavenging of the skin extract was higher (75.42%) than that of the pulp extract (53.31%). In Algeria, on the other hand, a percentage of 68.98% was reported by Ouchemoukh et al.,(2017).

CONCLUSION

The fig is one of the oldest domesticated fruit species, generally growing in hot, dry climatic zones. The fruits of fig trees are widely used. They are an excellent source of minerals, vitamins, amino acids, crude fiber and phenolic compounds. Figs have been traditionally used for their medicinal virtues as a remedy for many cardiovascular and respiratory diseases, as an antispasmodic, anti-inflammatory and expectorant, and as a hemorrhoidal remedy since ancient times.Today, the food industry attaches great importance to this fruit for its various uses. It is transformed into jams and marmalades, as well as juices. Fig quality is of prime importance to the consumer, including organoleptic quality (ripeness, firmness, odor and flavor), nutritional quality (protein, fiber, sugar, amino acid, mineral and vitamin content) and phytochemical composition (polyphenol content and various antioxidants).In general, all our research has confirmed the benefits of eating this fruit for its nutritional value and antioxidant content.Finally, as a perspective, we wish to carry out an experimental study on the different varieties of figs (Ficus carica) existing in our region "Jijel" in :

√ Evaluating physicochemical, nutritional and phytochemical quality, in particular antioxidant activity.
√ In vivo tests to better assess the antioxidant activity of fresh figs.
√ Quantification of other antioxidant elements to predict their effect on health.

√ It would be interesting to carry out sensory and microbiological analyses of fresh figs.

√ Monitor color and odor changes during storage of fresh figs.

REFERENCES

Abbas, T., Khatoon, S., Alam, R., Hussain, B., Hussain, Z., Gonzalez, M. Y., Abbas, Y., Ali, N., & Hussain, N. (2016). A Physico-Chemical study of different Fig (Ficus Carica L.) varieties in Haramosh valley, Gilgit-Pakistan. International Journal of Environment, Agriculture and Biotechnology, 1(3), 517-525.

Abd-Eltawab, S., & Ebid, W. (2019). Production and Evaluation of Stirred Synbiotic Fermented Milk Fortified with Fig Fruit (Ficus carica L.). Egyptian Journal of Food Science, 47(2), 201-212.

Abou-Farrag, H.T., Abdel-Nabey, A.A., Abou-Gharbia, H.A., & Osman, H.O. A.(2013). Physicochemical and Technological Studies on Some Local Egyptian Varieties of Fig (Ficus carica L.). Alexandria Science Exchange Journal, 34(2), 189-203.

Afriyanti, M., Mulyono, A., Basuki, J., & Sukaryani, S. (2018). Improving The Quality of Fig (Ficus carica. L) Processing from Posdaya Lancar Barokah Pokoh Kidul Wonogiri. In International Conference on Applied Science and Engineering (ICASE 2018). Atlantis Press. In food and agricultural industries, 96,15-628.

Ahmad, S., Bhatti, F. R., Khaliq, F. H., Irshad, S., & Madni, A. (2013). A review on the prosperous phytochemical and pharmacological effects of Ficus carica. International Journal of Bioassays, 2, 843-849.

Ait Haddou, L., Blenzar, A., Messaoudi, Z., Van Damme, P., Boutkhil, S., & Boukdame, A. (2014). Effect of cultivar, pretreatment and drying technique on some physicochemical parameters of dried figs from seven local cultivars of the fig tree (Ficus carica L.) in Morocco. European Journal of Scientific Research, 121(4), 336-346.

Åkesson, M. T., & di Caracalla, V. D. T. (2010). Draft Maximum Levels for Total Aflatoxins in Dried Figs, 11, 5-27.

Aksoy, U., Hakerlerler, H., Anac, D., & Duzbastilar, M. (1987). Soil properties and mineral content of Sarilop fig orchards in Germencik and relationship between mineral nutrients and yield and fruit quality parameters (Turkish).

Triangle Research and Developing Center, Bornova, 34.

Al Askari, G., Kahouadji, A., Khedid, K., Charof, R., & Mennane, Z. (2012). Physicochemical and microbiological characterizations of dried fig sampled from Rabat- Salé, Temara and Casablanca markets. Laboratory technologies, 7(26).

Alaa H., Ibrahim S. and Khalifa A., (2015). Improve sensory, quality and textural properties of fermented camel's milk by fortified with diatary fiber. Journal of American science. 11(3), 42-54.

Aljane, F. (2018). Evaluation of phenolic compounds and antioxidant activities of figs (Ficus carica L.). Bari-Chania-Montpellier-Zaragoza.

Aljane, F., & Ferchichi, A. (2009). Postharvest chemical properties and mineral contents of some fig (Ficus carica L.) cultivars in Tunisia. Journal of Food Agriculture and Environment 7 (2), 209-212.

Aljane, F., & Sdiri, N. (2014). Phytochemical characteristics as affected by fruit skin color of some fig (Ficus carica L.) accessions from southeastern Tunisia. Journal of New Sciences, 35, 10-139.

Aljane, F., Toumi, I., & Ferchichi, A. (2007). HPLC determination of sugars and atomic absorption analysis of mineral salts in fresh figs of Tunisian cultivars. African Journal of Biotechnology, 6(5), 599-602.

Amaro, L. F., Soares, M. T., Pinho, C., Almeida, I. F., Ferreira, I. M. P. L. V. O., & Pinho, O. (2012). Influence of cultivar and storage conditions in anthocyanin content and radical- scavenging activity of strawberry jams. World Academy of Science, Engineering and Technology, 69.

Amić, D., Davidović-Amić, D., Beslo, D., & Trinnajstić, N. (2003). Structure-Radical scavenging activity relationships of flavonoids. Croatica Chemica Acta, 76 (1), 55-61.

Anderson, C.M., Hallberg, A., & Hogberg, T.(1996). Advances in the development of pharmaceutical antioxidants. In Advances in Drug Research, 28, 65-180.

Anonymous (2005). documents Algériens, Série économique: agriculture le figuier et l'exportation des figues en Algérie. n°67; March 10, 1950.

Arpaci, S. (2015). An overview on fig production and research and development in Turkey. In International Symposium on Fig, 1173, 57-62.

Arumugam, P., Haritha , M., Keerthana, R., Vijayalakshmi ,M., Saraswathi, K. (2018).comparative antioxidant and antimicrobical activites of peel and pulp of fruits of ficus carica L, 7(15), 1176-1195.

Arvaniti, O. S., Samaras, Y., Gatidou, G., Thomaidis, N. S., & Stasinakis, A. S. (2019). Review on fresh and dried figs: Chemical analysis and occurrence of phytochemical compounds, antioxidant capacity and health effects. Food Research International, 119, 244-267.

Assaf, R. (2001). Local variety selection and cultivation techniques for fig trees in Israel. Fruits, 56(2), 101-121.

Awatef, E., Fethi, M. R., Nizar, C., Afraa, R., Belgacem, L., Leila, B. Y., & Ali, F. (2009). Evaluation of the total sugar composition of three local fig cultivars before and after drying. Evaluation, 15, 16-17.

Babazadeh Darazi, B. (2011). Morphological and pomological characteristics of fig (Ficus carica L.) cultivars from Varamin, Iran. African Journal of Biotechnology, 10 (82), 19096-19105.

Badgujar, S. B., Patel, V. V., Bandivdekar, A. H., & Mahajan, R. T. (2014). Traditional uses, phytochemistry and pharmacology of Ficus carica: A review. Pharmaceutical biology, 52(11), 1487-1503.

Bakshi, D. N.G., Sensarma, P., & Pal, D. C. (1999). A lexicon of medicinal plants in India. Naya Prakash, Calcutta, 424-425.

Balasundram, N., Sundram, K., & Sammam, S. (2006). Phenolic compounds in plants and agri- indistrual by-products: Antioxidant activity, occurrence, and potential uses. Food Chemistry, 99, 191-203.

Bayer, E., Buttler, K.P., Finkenzeller, X., & Grau, J. (2005). Guide to Mediterranean flora. édition Tec et Doc,Lavoisier Paris.12-13.

Bedioul, A., & Alouane, L. (2010). Traditional plants and food. Tunisie, 22-23.

Belitz, H.D., Grosch, W., & Schieberle, P. (2009). Fruits and Fruit Products. Food Chemistry. Springer Edition, 807-861.

Bidri, M. (2018). Early ripening of figs by olive oil: Involvement of the ethylene signaling pathway. Revue Marocaine des Sciences Agronomiques et Vétérinaires, 6(4), 489-493.

Bingham, S. A., Day, N. E., Luben, R., Ferrari, P., Slimani, N., Norat, T., & Tjφnneland, A. (2003). Dietary fibre in food and protection against colorectal cancer in the European Prospective Investigation into Cancer and Nutrition (EPIC): an observational study. The lancet, 361(9368), 1496-1501.

Bioavailability and bioefficacy of polyphenols in humans. American Journal of Clinical Nutrition, 81, 230S-242S.

Board, B. W. (1987). Quality control in the fruit and vegetable processing industry.

FAO Study: Food and Nutrition (FAO) fre notas 39, 70p .
Borel, P., Drai, J., Faure, H., Fayol, V., Galabert, C., Laromiguière, M., & Le Moël, G. (2005). Recent data on carotenoid absorption and catabolism. In Annales de Biologie Clinique, 63, 165-177.

Bradford, M.M. (1976). A rapid and sensitive method for the quantification of microgram quantities of protein utilizing the principle of protein-dye binding. Analytical Biochemistry, (72), 248-254.

Brien, J., & Hardy, T.S. (2002). Fig growing in NSW, Agfact H3.1.19, first edition Order N°H3.1.19 Agdex 219. Edited by Ann Munroe, 1-8.

Bruneton, J. (1999). "Pharmacognosy." Phytochemistry. Plantes medicinales, published by Tec-Doc, Paris.

Budić, L. I., & Lovrić, T. (2002). "Identification of phenolic acids and changes in their content during fermentation and ageing of white wines Pošip and Rukatac. Food Technology and Biotechnology, 40(3), 221.

Caliskan, O., & Polat, A.A. (2008). Fruit characteristics of fig cultivars and genotypes grown in Turkey. Scientia Horticulturae, 115, 360-367.

Çalişkan, O., & Polat, A.A. (2011). Phytochemical and antioxidant properties of selected fig (Ficus carica L.) accessions from the eastern Mediterranean region of Turkey. Scientia Horticulturae, 128(4), 473-478.

Caliskan, O., & Polat, A.A. (2012). Morphological diversity among fig (Ficus carica L.) accessions sampled from the eastern Mediterranean region of Turkey. Turkish Journal of Agriculture and Forestry, 36, 179-193.

Cao, G., Sofic, E., & Prior, R. L.(1997). Antioxidant and prooxidant behavior of flavonoids: structure-activity relationships. Free Radical Biology & Medicine, 22(5), 749-760.

Caraglio (2005). Secrets of the fig tree. Nature outings: from autumn to winter.

Chahidi, B., El-Otmani, M., Jacquemond, C., Tijane, M. H., El-Mousadik, A., Srairi, I., & Luro,

F. (2008). Use of morphological and physiological traits and molecular markers to assess the genetic diversity of three clementine cultivars. Comptes Rendus Biologies, 331(1), 1-12.

Chawla, A., Kaur, R., & Sharma, A. K. (2012). Ficus carica Linn.: A review on its pharmacognostic, phytochemical and pharmacological aspects. International Journal of Pharmaceutical and Phytopharmacological Research, 1(4), 215-232.

Cheikh,T. M. (2006). Study of the phytochemistry and biological activities of some plants used in the traditional treatment of dysmenorrhea in Mali. PhD thesis, 175.

Chouaki, S., Bessedik, F., Chebouti, A., Maamri, F., Oumata, S., Kheldoun, S., Hamana,M ., Douzene,M ., Bellah, M., & Kheldoun, A. (2006). Second national report on the status of plant genetic resources. Institut National De La Recherche Agronomique, 91P.

Condit, I. (1955). Fig varieties: a monograph. Hilgardia, 23(11), 323-538. Couplan, F. (1998). Guide nutritionnel des plantes. ED. Sophie Daguin.

Crisosto, C. H., Bremer, V., Ferguson, L., & Crisosto, G. M. (2010). Evaluating quality attributes of four fresh fig (Ficus carica L.) cultivars harvested at two maturity stages. HortScience, 45(4), 707-710.

Crisosto, H., Ferguson, L., Bremer, V., Stover, E., & Colelli, G. (2011). Fig (Ficus carica L.). In Postharvest biology and technology of tropical and subtropical fruits 134-160, Woodhead Publishing.

Darjazi, B.B. (2011). Morphological and pomological characteristics of fig (Ficus carica L.) cultivars from Varamin, Iran. African Journal of Biotechnology, 10 (82), 19096-19105.

Del Carmen Villalobos, M., Serradilla, M. J., Martín, A., Ruiz-Moyano, S., Pereira, C., & de Guía Córdoba, M. (2016). Synergism of defatted soybean meal extract and modified atmosphere packaging to preserve the quality of figs (Ficus carica L.). Postharvest Biology and Technology, 111, 264-273.

Del Carmen Villalobos, M., Serradilla, M. J., Martín, A., Ruiz-Moyano, S., Pereira, C., & de Guía Córdoba, M. (2014). Use of equilibrium modified atmosphere packaging for preservation of 'San Antonio'and 'Banane'breba crops (Ficus carica L.). Postharvest biology and technology, 98, 14-22.

Del Caro, A., & Piga, A. (2008). Polyphenol composition of peel and pulp of two Italian fresh fig fruits cultivars (Ficus carica L.). European Food Research and Technology, 226(4), 715-719.

Derbel, S., & Ghedira, K. (2005). Phytonutrients and their impact on health. Phytothérapie, 3(1), 28-34.

DSA. (2017). (Direction des Services Agricole). Béjaia.
Dubois, M., Gilles, K.A., Hamilton, J.K., Rebers, P.A., & Smith, F. (1956). Colorimetric method for determination of sugars and related substances. Analytical Chemistry, 28(3), 350- 356.

Dueñas, M., Pérez-Alonso, J. J., Santos-Buelga, C., & Escribano-Bailón, T. (2008). Anthocyanin composition in fig (Ficus carica L.). Journal of Food Composition and Analysis, 21(2), 107-115.

El-Agamey, A., Lowe, G. M., McGarvey, D. J., Mortensen, A., Phillip, D. M., Truscott, T. G., & Young, A. J. (2004). Carotenoid radical chemistry and antioxidant/pro-oxidant properties. Archives of biochemistry and biophysics, 430(1), 37-48.

Ercisli, S., Tosun, M., Karlidag, H., Dzubur, A., Hadziabulic, S., & Aliman, Y. (2012). Color and antioxidant characteristics of some fresh fig (Ficus carica L.) genotypes from Northeastern Turkey. Plant Foods for Human Nutrition, 67(3), 271-276.

Erlund, I. (2004). Review of the flavonoids quercetin, hesperetin, and naringenin. Dietary sources, bioactivities, bioavailability, and epidemiology. Nutrition research, 24(10), 851-874.

Faleh, E., Oliveira, A. P., Valentão, P., Ferchichi, A., Silva, B. M., & Andrade, P. B. (2012). Influence of Tunisian Ficus carica fruit variability in phenolic profiles and in vitro radical scavenging potential. Revista Brasileira de Farmacognosia, 22(6), 1282-1289.

FAO. (2016). (Food and Agriculture Organization of the United Nations).

Farahnaky, A., Ansari, S., & Majzoobi, M. (2009). Effect of glycerol on the moisture sorption isotherms of figs. Journal of Food Engineering, 93, 468-473.

Favier, J.C., Ireland-Ripert, J., Laussucq, C., & Feinberg, M. (1993). Répertoire général des aliments. Tome 3: table de composition des fruits exotiques, fruits de cueillette d'Afrique. Editions Office de la recherche scientifique et technique outre-mer (ORSTOM) and Tech & Doc, L'Institut national de la recherche agronomique (INRA), 31-34.

Ferradji, A., Chabour, H., & Malek, A. (2011). Solar drying of figs: Heat balance and desorption isotherm. Revue des énergies renouvelables, 14(4), 717-726.

Ferreiro-Vera, C., Mata-Granados, J. M., Gómez, J. Q., & de Castro, M. L. (2011). On-line coupling of automatic solid-phase extraction and HPLC for determination of carotenoids in serum. Talanta, 85(4), 1842-1847.

Flaishman, M. A., Rodov, V., & Stover, E. (2008). The fig: botany, horticulture, and breeding , 34,119-125.

Formica, J. V., & Regelson, W. (1995). Review of the biology of quercetin and related bioflavonoids. Food and chemical toxicology, 33(12), 1061-1080.

Freiman, Z.E, Rodov, V., Yablovitz, Z., Horev, B., & Flaishman, M.A. (2012). Pre-harvest application of 1-methylcyclopropene inhibits ripening and improves storage quality of 'brown turkey' (Ficus carica L.) sprigs. Scientia Horticulturae, 138, 266-272.

Gabriele, D., Migliori, M., & De Cindio, B. (2010). Innovation in fig syrup production process: a rheological approach. International journal of food science & technology, 45(9), 1947- 1955.

Gardès, A. M., Bonnefont, R. D., & Abedinzadeh, Z. (2003). Reactive oxygen species: how oxygen can become toxic. L'actualité chimiques, 91-96.

Garrone, B. (1998). The fig tree. Les écologistes de l'Euzière. 2nd edition, Presses du Midi, Montpelier, 111 .

Gaussen, H., Leroy, J.F., & Ozenda, P. (1982). Précis de botanique, tome II: végétaux supérieure. Masson, grenadier. Transfer Génétique en Agriculture, 105, 558-560.

Ghedira, K. (2005). Flavonoids: structure, biological properties, prophylactic role and therapeutic uses. Phytothérapie, 3(4), 162-169.

Golubev, V.N., Pilipenko, L.N., & Kakhniashvili, T.A. (1987). Fractionation and composition of the carbohydrates of Ficus carica. Plenum Publishing Corporation, 631-634.

Guttnneau, G. (1992). Connaitre et reconnaitre la flore et la végétation méditerranéenne, Ed.Ouest-France, 331.
Guvenc, M. E. (2009). Analysis of fatty acid and some lipophilic vitamins found in the fruits of the (Ficus carica) variety picked from the Adiyaman district. Research Journal of Biological Sciences, 4 (3), 320-323.

Haesslein, D., & Oreiller, S. (2008). Fresh or dried, the fig is revealed. Filière Nutrition et diététique. Haute Ecole de Santé Genève, 1-4.

Handique, J.G., & Baruah, J.B. (2002). Polyphenolic compounds an overview. Reactive &Functional Polymers, 52, 163-188.

Harborne, J.B., Williams, C.A.(2000). Advances in flavonoid research since 1992.Phytochemistry, 55(6), 481-504.

Harzallah, A., Bhouri, A. M., Amri, Z., Soltana, H., & Hammami, M. (2016). Phytochemical content and antioxidant activity of different fruit parts juices of three figs (Ficus carica L.) varieties grown in Tunisia. Industrial Crops and Products, 83, 255-267.

Hertog, M. G., Hollman, P. C., Katan, M. B., & Kromhout, D. (1993). Intake of potentially anticarcinogenic flavonoids and their determinants in adults in The Netherlands. Nutrition and Cancer 20, 21-29.

Hollman, P.C., Hertog, M.G.L., & Katan, M.B. (1996). Analysis and health effects of flavonoids. Food Chemistry, 57, 43-46.

Hoxha, L., & Kongoli, R. (2016). Evaluation of antioxidant potential of albanian fig varieties "kraps zi" and "kraps bardhe" cultivated in the region of Tirana. Journal of Hygienic Engineering and Design, 16, 70-74.

Huang, W.Y., Cai, Y.Z., Corke, H., & Sun, M. (2010). Survey of antioxidant capacity and nutritional quality of selected edible and medicinal fruit plants in Hong Kong. Journal of Food Composition and Analysis, 23, 510-517.

Ibrahim, A. H., & Khalifa, S. A. (2015). Improve sensory quality and textural properties of fermented camel's milk by fortified with dietary fiber. Journal of American Science, 11(3), 42-54.

Imran, A., Jat, R. K., & Varnika, S. (2011). A review on traditional, pharmacological, pharmacognostic properties of Ficus carica (Anjir). International Research Journal of Pharmcy,12, 124-127.

ITAFV. (2003). Etude sectorielle de l'arboriculture fruitiére et de la viticulture en Algérie. Institut Technique de l'Arboriculture Fruitière et de la Vigne, 30 P.

James, I. F., & Kuipers, B. (2003). AD03F Fruit and vegetable preservation. Agromisa Foundation,94P.

Jean-Jacques, M. (1996). Les composes phenoliques des végétaux: quelles perspectives à la fin du XXeme siecle . Acta Botanica Gallica , 473-479.

Jiang, L., Shen, Z., Zheng, H., He, W., Deng, G., & Lu, H. (2013). Noninvasive evaluation of fructose, glucose, and sucrose contents in fig fruits during development using chlorophyll fluorescence and chemometrics. Journal of Agricultural Science and Technology, 15, 333-342.

Jokić, S., Mujić, I., Bucić-Kojić, A., Velić, D., Bilić, M., Planinić, M., & Lukinac, J. (2014). Influence of extraction type on the total phenolics, total flavonoids and total colour change of different varieties of fig extracts. Hrana u zdravlju i bolesti: znanstveno- stručni časopis za nutricionizam i dijetetiku, 3(2), 90-95.

Kakhniashvili, T. A., Kolesnik, A. A., Zherebin, Y. L., & Golubev, V. N. (1986). Liposoluble pigments of the fruit of Ficus carica. Chemistry of Natural Compounds, 22(4), 477-479.

Kjellberg, F., Doumesche, B., & Bronstein, J.(1988). Longevite of fig wasp (Blastophaga psenes). Proceeding of the Koninklijke Nederlandse Academie VanWetenshppen. Serie C, 122-171.
Kolesnik, A. A., kakhniashvili, T.A., Zherebin, Y. L., Golubev, V. N., & Pilipenko, L. N. (1987).

Lipids of the fruit of Ficus carica. Plenum Publishing Corporation Ukranie, 394-397.
Kong, J. M., Chia, L. S., Goh, N. K., Chia, T. F., & Brouillard, R. (2003). Analysis and biological activities of anthocyanins. Phytochemistry, 64(5), 923-933.

Lacaille-Dubois, M.A., & Wagner, H. (1996). Pharmacological significance of polyphenolic derivatives. Acta botanica gallica, 143(6), 555-562.

Lacroix, M. (2008). Qualitative and quantitative variations in dairy protein intake in animals and humans: metabolic implications. Thesis to obtain the degree of Doctor of Agro Paris Tech.

Lahmadi, A., Filali, H., Samaki, H., Zaid , A., Aboudkhil , S., (2019). Phytochemical screening, antioxidant activity and inhibitory potential of Ficus carica and Olea europaea leave. Bioinformation, 15(3), 226.

Lansky, E. P., Paavilainen, H. M., Pawlus, A. D., & Newman, R. A. (2008). Ficus spp.(fig): Ethnobotany and potential as anticancer and anti-inflammatory agents. Journal of Ethnopharmacology, 119, 195-213.

Lansky, E.P., & Paavilainen, H.M. (2011). Traditional herbal medicines for modern times: Figs the Genus Ficus. ED. Taylor and Francis Group, LLC Chemical Rubber Company Press, 357.

Lean, M. E., Noroozi, M., Kelly, I., Burns, J., Talwar, D., Sattar, N., & Crozier, A. (1999). Dietary flavonols protect diabetic human lymphocytes against oxidative damage to DNA. Diabetes, 48(1), 176-181.

Lee, C.Y., Shallenberger, R. S., & Vittum, M. T. (1970). Free sugars in fruits and vegetables.

Food Science and Technology, 1, 1-12.

Leroux, H., & Schuber, T. E. (1983). Pectin applications in the food industry. In: industries alimentaires et agricoles, 9,615-628.

Leroy , J.F. (1968). Tropical and subtropical fruits. Institut français de la recherche fruitière outre-mer. 1st edition, Presse universitaire de France,7-50.

Liao, M.L., & Seib, P.A. (1988). Chemistry of L-ascorbic acid related to foods. Food Chemistry, 30, 289-312.

Lien, E. J., Ren, S., Bui, H. H., & Wang, R. (1999). Quantitative structure-activity relationship analysis of phenolic antioxidants. Free Radical Biology and Medicine, 26(3-4), 285- 294.

Lim, T.K.(2012). Edible medicinal and non-medicinal plants: Ficus carica. Moraceae. Edition Springer Sciences Media B.V. Fruits, 3, 362-376.

Lopez Camelo, A. F. (2007). Manual for the preparation and sale of fruit and vegetables. From field to market. Bulletin of the Agricultural Services of the Food and Agriculture Organization of the United Nations, Rome, 151, 39-182.

Lugasi, A., Hóvári, V., Sági, K., &Bíró, L. (2003). The role of antioxidant phytonutrients in the prevention of diseases. Acta Biologica Szegediensis 47(1-4), 119-125.

MADR. (2005).Statistiques agricoles, série A et B,Ministére de l'Agriculture, du Développement Rural (Algiers).
MADR. (2012). Ministry of Agriculture and Rural Development .Service of the Statistics (Algiers, Algeria).

Mahmoudi, S., Khali, M., Benkhaled, A., Benamirouche, K., & Baiti, I. (2016). Phenolic and flavonoid contents, antioxidant and antimicrobial activities of leaf extracts from ten Algerian Ficus carica L. varieties. Asian Pacific Journal of Tropical Biomedicine, 6(3), 239-245.

Mahmoudi, S., Khali, M., Benkhaled, A., Boucetta, I., Dahmani, Y., Attallah, Z., & Belbraouet,

S. (2018). Fresh figs (Ficus carica L.): Pomological characteristics, nutritional value, and phytochemical properties. European Journal of Horticultural Science, 83(2), 104- 113.

Manach, C., Scalbert, A., Morand, C., Rémésy, C., & Jiménez, L. (2004). Polyphenols: food sources and bioavailability. The American journal of clinical nutrition, 79(5), 727-747.

Manach, C., Williamson, G., Morand, C., Scalbert, A., & Rémésy, C. (2005). Bioavailability and bioefficacy of polyphenols in humans. I. Review of 97 bioavailability studies. The American journal of clinical nutrition, 81(1), 230S-242S.

Marei, N., & Crane, J. C. (1971). Growth and respiratory response of fig (Ficus carica L. cv. Mission) fruits to ethylene. Plant Physiology, 48(3), 249-254. Marinova, D., Ribarova, F., & Atanassova, M. (2005). Total phenolics and total flavonoids in bulgarian fruits and vegetables. Journal of the University of Chemical Technology and Metallurgy, 40 (3), 255-260.

Mars, M. (2001). Fig (Ficus carica L.) genetic resources and breeding. In II International Symposium on Fig, 605 , 19-27.

Mawa, S., Husain, K., & Jantan, I. (2013). Ficus carica L.(Moraceae): phytochemistry, traditional uses and biological activities. Evidence-Based Complementary and Alternative Medicine, 2013, 1-8.

Mehraj, H., Sikder, R. K., Haider, M. N., Hussain, M. S., & Jamal Uddin, A. F. M. (2013). Fig (Ficus carica L.): a new fruit crop in Bangladesh. International Journal of Business, Social and Scientific Research, 1(1), 1-5.

Mendoza-Castillo, V. M., Pineda-Pineda, J., Vargas-Canales, J. M., & Hernández-Arguello, E. (2019). Nutrition of fig (Ficus carica L.) under hydroponics and greenhouse conditions. Journal of Plant Nutrition, 42(11-12), 1350-1365.

Meziant, L., Saci, F., Bachir Bey, M., & Louaileche, H. (2015). Varietal influence on biological properties of Algerian light figs (Ficus carica L.). International Jourenal. Bioinform Biomed Engineering, 1, 237-243.

Miguel, M.G. (2011). Anthocyanins: Antioxidant and/or anti-inflammatory activities. Journal of Applied Pharmaceutical Science, 1(6), 7-15.

Montealegre, R. R., Peces, R. R., Vozmediano, J. C., Gascueña, J. M., & Romero, E. G. (2006). Phenolic compounds in skins and seeds of ten grape Vitis vinifera varieties grown in a warm climate. Journal of Food Composition and Analysis, 19(6-7), 687-693.

Naczk, M., & Shahidi, F. (2004). Extraction and analysis of phenolics in food. Journal of chromatography A, 1054(1-2), 95-111.

Nijveldt, R. J., Van Nood, E. L. S., Van Hoorn, D. E., Boelens, P. G., Van Norren, K., & Van Leeuwen, P. A. (2001). Flavonoids: a review of probable mechanisms of action and potential applications. clinical nutrition, 74, 418-425.

Oliveira, A.P., Valentão, P., Pereira, J.A., Silva, B.M., Tavares, F., & Andrade, P.B. (2009). Ficus carica L.: Metabolic and biological screening. Food and Chemical Toxicology, (47), 2841-2846.

Omondi Owino, W., Nakano, R., Kubo, Y., & Inaba, A. (2004). Alterations in cell wall polysaccharides during ripening in distinct anatomical tissue regions of the fig (Ficus carica L.) fruit. Postharvest Biology and Technology, 32(1), 67-77.

Osawa, T., & Namiki, M. (1981). A novel type of antioxidant isolated from leaf wax of Eucalyptus leaves. Agricultural and biological chemistry, 45(3), 735-739.

Ouaouich, A., & Chimi, H. (2005). Guide du sécheur de figues. 1st edition. United Nations Industrial Development Organization, Morocco, 10, 28.

Ouchemoukh, N., Ouchemoukh, S., Meziant, N., Idiri, Y., Hernanz, D., Stinco, C. M., & Luis, J. (2017). Bioactive metabolites involved in the antioxidant, anticancer and anticalpain activities of Ficus carica L., Ceratonia siliqua L. and Quercus ilex L. extracts. Industrial Crops and Products, 95, 6-17.

Ouchemoukh, S., Hachoud, S., Boudraham, H., Mokrani, A., & Louaileche, H. (2012). Antioxidant activities of some dried fruits consumed in Algeria. LWT-Food Science and Technology, 49(2), 329-332.

Oukabli, A. (2003). Le Figuier: un patrimoine génétique diversifié à exploiter. L'Institut national de la recherche agronomique (INRA), Transfert de technologie en agriculture, 106(4).

Ozer, B.K., & Derici, B. (1998). A research on the relationship between aflatoxin and ochratoxin a formation and plant nutrients. Acta Hortic, 480, 199-

206.

Pande, G., & Akoh, C.C. (2010). Organic acids, antioxidant capacity, phenolic content and lipid characterisation of Georgia-grown underutilized fruit crops. Journal of Food Chemistry 120, 1067-1075.

Patil, A. P., Kad, V. P., & Shelar, S. D. (2017). Studies on Preparation of Fig Jam Without Preservative. Journal of Agriculture Research and Technology, 42(3), 204.

Pereira, C., López Corrales, M., Martín, A., Villalobos, M.D.C., Córdoba, M.D.G., & Serradilla, M. J.(2017). Physicochemical and nutritional characterization of brebas for fresh consumption from nine fig varieties (Ficus carica L.) grown in Extremadura (Spain). Journal of Food Quality, 2017, 1-12.

Pereira, C., Martín, A., López-Corrales, M ., de Guía Córdoba, M ., Galván, A. I., & Serradilla,

M. J. (2020). Evaluation of the Physicochemical and Sensory Characteristics of Di erent Fig Cultivars for the Fresh Fruit Market. Foods, 9(619), 1-16.

Pereira, M. A., O'Reilly, E., Augustsson, K., Fraser, G. E., Goldbourt, U., Heitmann, B. L., & Spiegelman, D. (2004). Dietary fiber and risk of coronary heart disease: a pooled analysis of cohort studies. Archives of internal medicine, 164(4), 370-376.

Petkova, N., Ivanov, I., & Denev, P. (2019). Changes in phytochemical compounds and antioxidant potential of fresh, frozen, and processed figs (Ficus carica L.). International Food Research Journal, 26(6), 1881-1888.

Preedy, V. R., & Watson, R. R. (2014). The mediterranean diet: an evidence-based approach. Academic press, 629-637.
Prior, R. L., Wu, X., & Schaich, K. (2005). Standardized methods for the determination of antioxidant capacity and phenolics in foods and dietary supplements. Journal of agricultural and food chemistry, 53(10), 4290-4302.

Puoci, F., Iemma, F., Spizzirri, U. G., Restuccia, D., Pezzi, V., Sirianni, R., & Picci, N. (2011). Antioxidant activity of a Mediterranean food product: "fig syrup". Nutrients, 3(3), 317- 329.

Quideau, S., Deffieux, D., Douat-Casassus, C., & Pouysegu, L. (2011). Plant polyphenols: chemical properties, biological activities, and synthesis. Angewandte Chemie International Edition, 50(3), 586-621.

Rameau, J. C., Mansion, D., & Dumé, G. (2008). French forest flora: illustrated ecological guide. Mediterranean region 3, French private forest.

Ramulu, P., & Rao, P. U. (2003). Total, insoluble and soluble dietary fiber contents of Indian fruits. Journal of food composition and analysis, 16(6), 677-685.

Rao, A. V., & Rao, L. G. (2007). Carotenoids and human health. Pharmacological research, 55(3), 207-216.

Rice-Evans, C.A., Miller, N.J., Bolwell, P.G., Bramley, P.M., & Pridham, J.B. (1995). The relative antioxidant activities of plant-derived polyphenolic flavonoids. Free Radical Research, 22, 375-383.

Robards, K. (2003). Strategies for the determination of bioactive phenols in plants, fruit and vegetables. Journal of chromatography A, 1000(1-2), 657-691.

Rogez, H., Pompeu, D. R., Akwie, S. N. T., & Larondelle, Y. (2011). Sigmoidal kinetics of anthocyanin accumulation during fruit ripening: a comparison between açai fruits (Euterpe oleracea) and other anthocyanin-rich fruits. Journal of food Composition and Analysis, 24(6), 796-800.

Sandhu, K.S., Singh, M., & Ahluwalia, P. (2001). Studies onprocessing of guava into pulp and guava leather. Journal of Food Science and Technology, 38, 622-624.

Sass-Kiss, A., Kiss, J., Milotay, P., Kerek, M. M., & Toth-Markus, M. (2005). Differences in anthocyanin and carotenoid content of fruits and vegetables. Food Research International, 38(8-9), 1023-1029.

Sekli-Belaidi, F. (2011). Functionalization of electrode surfaces by a poly (3, 4-ethylenedioxythiophene) PEDOT film for the elaboration of specific microsensors of ascorbic and uric acids: application to the study of antioxidant properties of blood serum (Doctoral dissertation, University of Toulouse, University Toulouse III-Paul Sabatier).

Scalbert, A., & Williamson, G. (2000). Dietary intake and bioavailability of polyphenols. The Journal of nutrition, 130(8), S2073-S2085.

Sengun, I. Y. (2013). Microbiological and chemical properties of fig vinegar produced in Turkey. African Journal of Microbiology Research, 7(20), 2332-2338.

Seyoum, A., Asres, K., & El-Fiky, F.K. (2006). Structure-radical scavenging activity relationships of flavonoids. Phytochemistry, 67(18), 2058-2070.

Shamin-Shazwan, K., Shahari, R., Amri, C. N. A. C., & Tajuddin, N. S. M. (2019). Figs (Ficus Carica L.): Cultivation Method and Production Based in Malaysia. Engineering Heritage Journal, 3(2), 06-08.

Silva, L. C., Harder, M. N., Arthur, P. B., Lima, R. B., Modlo, D. M., & Arthur, V.(2009). Physical-chemical characteristics of figs (Ficus carica) preready to submitted to ionizing radiation. International Nuclear Atlantic Conference, 41(26),1-9 .

Simoneliene, A., Treciokiene, E., Lukosiunaite, G., Vysniauskas, G., & Kasparaviciute, E.(2014). Rheology, technological and sensory characteristics of fortified drink products with fibers. Foodbalt, 294-297.

Simsek, M.(2009). Fruit performances of the selected fig types in Turkey. African Journal of Agricultural Research, 4 (11), 1260-1267.

Slatnar, A., Klancar, U., Stampar, F., & Veberic, R. (2011). Effect of drying of figs (Ficus carica L.) on the contents of sugars, organic acids, and phenolic compounds. Journal of Agricultural and Food Chemistry, 59(21), 11696-11702.

Solomon, A., Golubowicz, S., Yablowicz, Z., Grossman, S., Bergman, M., Gottlieb, H.E., Altman, A., Kerem, Z., & Flaishman, M.A. (2006). Antioxidant activities and anthocyanin content of fresh fruits of common fig (Ficus carica L.). Journal of Agricultural Food Chemistry, 54, 7717-7723.

Spigno, G., Tramelli, L., & De Faveri, D. M. (2007). Effects of extraction time, temperature and solvent on concentration and antioxidant activity of grape marc phenolics. Journal of food engineering, 81(1), 200-208.

Stalin, C., Dineshkumar, P., & Nithiyananthan, K.(2012). Evaluation of antidiabetic activity of methanolic leaf extract of Ficus Carica in alloxan-induced diabetic rats. Asian Journal of pharmaceutical and clinical Research, 5, 1-3.

Starr, F., Starr, K., & Loope, L. (2003). Ficus carica Edible fig Moraceae. Haleakala Field Station, Maui, Hawaii, 1-6.

Stover, E., Aradhya, M., Ferguson, L., & Crisosto, C. H. (2007). The fig: overview of an ancient fruit. HortScience, 42(5), 1083-1087.

Sun, B., Ricardo-da-Silva, J. M., & Spranger, I. (1998). Critical factors of vanillin assay for catechins and proanthocyanidins. Journal of agricultural and food chemistry, 46(10), 4267-4274.

Tabart, J., Kevers, C., Pincemail, J., Defraigne, J.O. & Dommes, J. (2010). Evaluation of spectrophotometric methods for antioxidant compound measurement in relation to total antioxidant capacity in beverages. Food Chemistry, 120, 607-614.

Tapiero, H., Townsend, D. M., & Tew, K. D. (2004). The role of carotenoids in the prevention of human pathologies. Biomedicine & Pharmacotherapy, 58(2), 100-110.

Taubert, D., Breitenbach, T., Lazar, A., Censarek, P., Harlfinger, S., Berkels, R., & Roesen, R. (2003). Reaction rate constants of superoxide scavenging by plant antioxidants. Free Radical Biology and Medicine, 35(12), 1599-1607.

Thaipong, K., Boonprakob, U., Crosby, K., Cisneros-Zevallos, L., & Hawkink Byrne, D. (2006). Comparison of ABTS, DPPH, FRAP, and ORAC assays for estimating antioxidant activity from guava fruit extracts. Journal of Food Composition and Analysis, 19(6-7), 669-675.

Tomas, B. F.A., & Clifford, M.N. (2000). Dietary hydroxyl benzoic derivativesnature, occurrence and dietary burden. Journal of the Science of Food and Agriculture, 80, 1024-1032.

Trad, M., Bourvellec, C., Gaaliche, B., Renard, C. M., & Mars, M. (2014). Nutritional compounds in figs from the southern Mediterranean region. International Journal of Food Properties, 17(3), 491-499.

Troszyńska, A., Estrella, I., López-Amóres, M. L., & Hernández, T. (2002). Antioxidant activity of pea (Pisum sativum L.) seed coat acetone extract. LWT-Food Science and Technology, 35(2), 158-164.

Tu Toit, R., Volsteedt, Y., & Apostolides, Z. (2001). Comparison of the antioxidant content of fruits, vegetables and teas measured as vitamin C equivalents. Toxicology, 166(1-2), 63- 69.

Valero, D., & Serrano, M.(2010). "Fruit ripening," in Postharvest Biology and Technology for Preserving Fruit Quality. Chemical Rubber Company, 4-47.

Valizadeh, M., Valdeyron, G., Kjellberg, F., & Ibrahim, M. (1987). Gene flow in the fig tree. Ficus carica: pollen dispersal in a dense stand. Acta Oecologica, 8 (22), 143-154.

Vandi, D., Nga, E. N., Betti, J. L., Loe, G. M. E., Ottou, P. B. M., Priso, R. J., & Mpondo, E. M. (2016). Contribution of the populations of the cities of Yaoundé and Douala to the knowledge of tannin and anthocyanin plants. Journal of Animal &Plant Sciences, 30(3), 4797- 4814.

Veberic, R., & Mikulic-Petkovsek ., M.(2016). Phytochemical composition of common fig (Ficus carica L.) cultivars. In Nutritional composition of fruit cultivars. Academic Press, 235-255.
Veberic, R., Colaric, M., & Stampar, F. (2008). Phenolic acids and flavonoids of fig fruit (Ficus carica L.) in the northern Mediterranean region. Food chemistry, 106(1), 153-157.

Ven, B., & Mann, J. (2004). Cereal grains, legumes and diabetes. European Journal of Clinical Nutrition, 58(11), 1443-1461.

Vidaud, J. (1997). Le figuier. Monographie de CTIFL (Centre international interprofessionnel des fruits et légumes), 267.

Vinson, J. A.(1999). The functional food properties of figs. Cereal foods world, 44(2), 82-87.
Vinson, J. A., Dabbagh, Y. A., Serry, M. M., & Jang, J. (1995). Plant flavonoids, especially tea flavonols, are powerful antioxidants using an in vitro oxidation model for heart disease. Journal of Agricultural and Food Chemistry, 43(11), 2800-2802.

Waghmare, R. B., & Annapure, U. S. (2018). Integrated effect of radiation processing and modified atmosphere packaging (MAP) on shelf life of fresh fig. Journal of food science and technology, 55(6), 1993-2002.

Walali, L., Skiredj, A., & Alattir, H. (2003). Fiches Techniques: L'amandier, l'olivier, le figuier, the pomegranate tree. Bulletin de Transfert de Technologie en Agriculture, 105, 3-4.

Yemm, E.W., & Cocking, E.C. (1955). The determination of amino acids with ninhydrin. Analyst, 80, 209-213.

Young, A.J., & Lowe, G.M. (2001). Antioxidant and prooxidant properties of carotenoids. Archives of Biochemistry and biophysics, 385(1), 20-27.

Zhang, K., & Jiang, R. (2006). Pharmacological study of Ficus carica. Zhongguo LinchuangKangfu, 10, 226-8

Zheng, W., & Wang, S.Y.(2003). Oxygen radical absorbing capacity of phenolics in Blueberries, Cranberries, Chokeberries, and Lingonberries. Journal of Agriculture Food Chemistry, 51, 502-509.

Zimmer, N., & Cordesse, R. (1996). Influence of tannins on the nutritional value of ruminant feeds. Productions animales, 9(3), 167-179.

SUMMARY

The fig (Ficus carica L.) is the fruit of the fig tree, a member of the Moraceae family, which is the emblem of the Mediterranean basin, where it has been cultivated for thousands of years. Mediterranean countries are the main producers of figs. Algeria produces 12.54% of the world's total fig crop. This fruit is a rich source of fiber, trace elements, antioxidant polyphenols, proteins, sugars, organic acids and volatile compounds that provide a characteristically pleasant aroma. Recently, a great deal of research has focused on the nutritional, physicochemical and pharmacological quality of this fruit. The similar composition of fig fruit is reported by various researchers as 77.5 to 86.8% moisture, 2.53 to 3.08% ash, 9.453 to 26.016 (g / 100 g) carbohydrates, 4.4 to 6.7% protein and 4 to 7.4% fiber by fresh weight. Phenolic compounds are a group of secondary metabolites that contribute not only to the fruit's sweet, bitter and astringent taste, but also to its aroma. Figs are rich in glycosidic flavonols, anthocyanins and other polyphenols, which increase antioxidant activity and consequently their consumption can have beneficial effects on health.

Key words: Ficus carica L., nutritional quality, physicochemical quality, polyphenols, antioxidant activity.

More
Books!

Printed by Books on Demand GmbH, Norderstedt / Germany